(No Model.)
N. TESLA
DYNAMO ELECTRI
No. 406,968.
D0457756
Fig. 1
Fig. 2

LIGHTNING STRIKES

JOHN F. WASIK

LIGHTNING STRIKES

TIMELESS LESSONS IN CREATIVITY FROM THE LIFE AND WORK OF NIKOLA TESLA

An Imprint of Sterling Publishing Co., Inc.
1166 Avenue of the Americas
New York, NY 10036

ISBN 978-1-4549-1768-7

Distributed in Canada by Sterling Publishing Co., Inc.
c/o Canadian Manda Group, 664 Annette Street
Toronto, Ontario, Canada M6S 2C8
Distributed in the United Kingdom by GMC Distribution Services
Castle Place, 166 High Street, Lewes, East Sussex, England BN7 1XU
Distributed in Australia by NewSouth Books
45 Beach Street, Coogee, NSW 2034, Australia

For information about custom editions, special sales, and premium and corporate purchases, please contact Sterling Special Sales at 800-805-5489 or specialsales@sterlingpublishing.com.

Manufactured in the United States of America

2 4 6 8 10 9 7 5 3 1

www.sterlingpublishing.com

Frontispiece: This multiple-exposure image created by *Century Magazine* photographer Dickenson Alley shows Tesla in his Colorado Springs laboratory with his massive "magnifying transmitter" generating 22-foot-long lightning bolts.

To Arthur Stanley Wasik

CONTENTS

Preface ix

Introduction: Why Tesla Still Electrifies Us 1

I. Hoover's Stakeout **9**
An Iconic Inventor Dies

TeslAction #1: Be Introspective 26

II. Catching Fire **29**
Tesla's Creative Inheritance

TeslAction #2: Be Unrelentingly Curious 46

III. Flashes of Light **49**
The Psychology of Tesla

TeslAction #3: Practice Visualization 66

IV. From Prague to Pittsburgh **69**
The Roving Electrician

TeslAction #4: Be Indispensable 92

V. The Wizard of Electricity **95**
Tesla's Lightning Show

TeslAction #5: Sell Your Ideas 116

VI. **POWER FROM THE EARTH AND SKY** **119**
Tesla's Wireless World System
TESLACTION #6: Zoom Out 138

VII. **WEATHERING THE STORM** **141**
The Fall of Wardenclyffe
TESLACTION #7: Be Resilient 156

VIII. **MAN FOR ALL SEASONS** **159**
Adapting to a New, Turbulent Century
TESLACTION #8: Reposition Yourself 184

IX. **A DETECTIVE STORY** **187**
Searching for the Elusive Death Ray
TESLACTION #9: Be Empathic 208

X. **THE ONCE AND FUTURE TESLA** **211**
The Wizard's Twenty-First-Century Legacy
TESLACTION #10: Be Chimeric 236

Epilogue *239*
References & Abbreviations *245*
Bibliographic Essay *249*
Acknowledgments *265*
Image Credits *267*
Index *269*
About the Author *276*

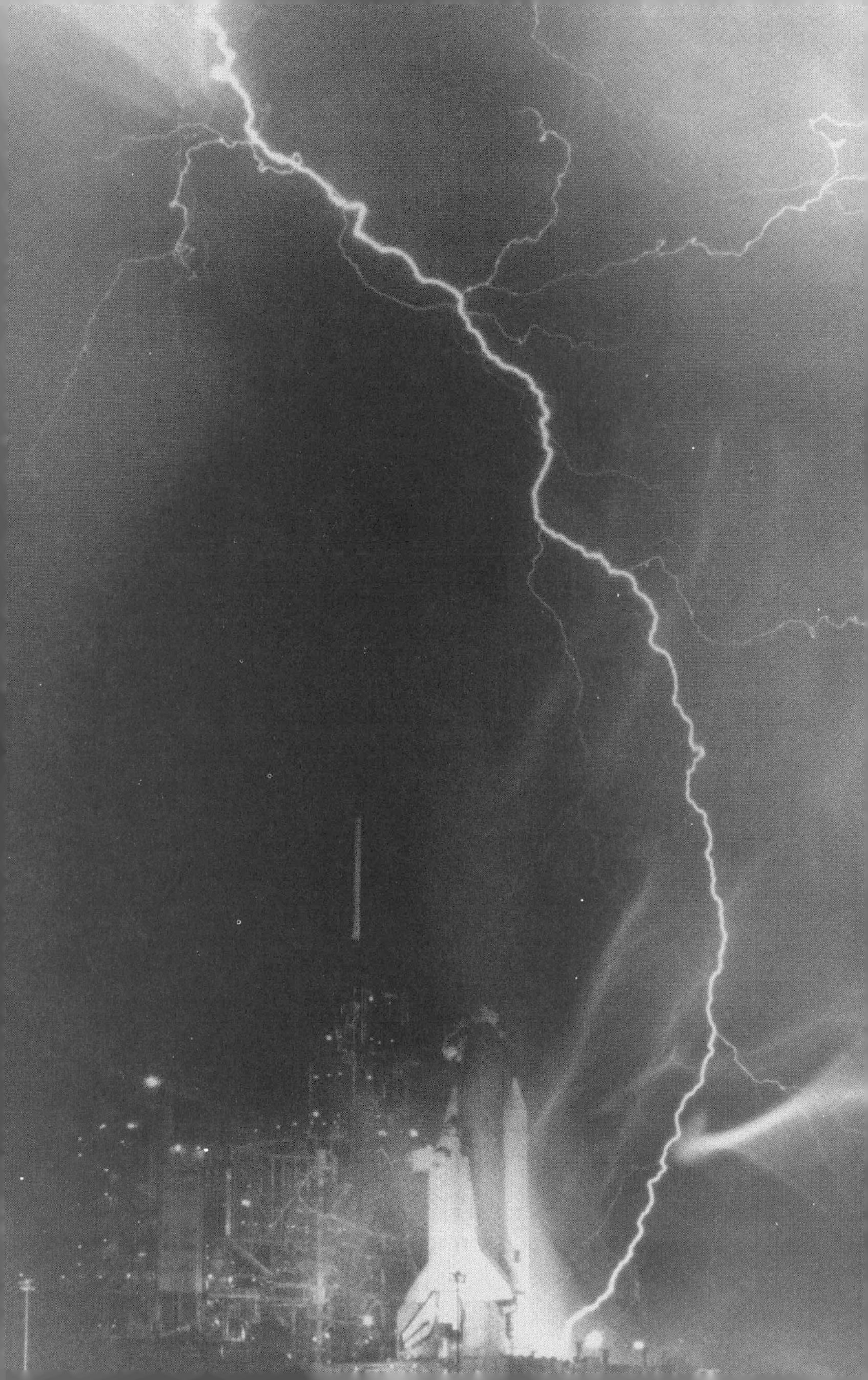

PREFACE

It began with a lightning storm. The lightning—bolts of pure energy like jagged, outstretched tendrils from the sky—touched the earth with such force it could destroy or illuminate thousands of cities if it ever could be captured, bottled, and rechanneled. That was how Nikola Tesla's life began, during a thunderstorm at the crack of midnight on July 10, 1856, in Smiljan, a village in present-day Croatia, on the military frontier of the Austro-Hungarian Empire.

I've always felt connected to this celestial energy, observing the dramatic thunderstorms of the American Midwest for nearly six decades. At first, the events were terrifying, but early on my mother gently told me that the flash and rumble were "just the angels bowling." With my mother's innocent alibi in my child's mind, from then on I was never afraid of the tempest, yet I always yearned for more details. How was it harnessed? How could it be our salvation and/or the source of our peril? How were the angels involved? Who was stacking the pins?

My father studied electronics in the navy in Washington toward the end of World War II in an attempt to thwart Hitler's V-2 program. When I was young, he put me in front of everything electronic and electromagnetic. I started with a crystal radio, then moved onto amplifiers, Van de Graaf generators, and Apple Inc.'s first attempt at a portable computer: the Apple® IIC.

My dad's workshop was littered with Edison-style wax cylinder Dictaphones, which were still somewhat in use in the late 1950s and early 1960s for office dictation, even though they represented the height of the 1870s technology boom. They contained motors and pulleys that my dad, a poor child of the Great

In a powerful evocation of Tesla's formidable legacy, an electrical storm at the Kennedy Space Center in Cape Canaveral, Florida, creates an eerie tapestry of light in the hours preceding the launch of the space shuttle *Challenger* in 1983.

Depression, believed could be salvaged for other, future uses. Although they were extremely simple machines, they still worked.

My boyhood job, other than to build electronic devices from kits, was to sort capacitors, resistors, bolts, and magnets into jars and know where everything was in my father's workshop, including every one of his hundreds of tools. Eventually I acquired my own soldering gun, which I bought with my own money. It was more precious than my bike or my baseball glove.

As I grew older, I started my own "laboratory" under the basement stairs. I was convinced that, during the space race, I could conceive of a great formula to reach the stars with my kid's chemistry set and all the electronic junk lying in the basement. Albert Einstein, Thomas Edison, and John Glenn were my heroes at the time. Once, when I was pretending to be floating outside of my capsule during a spacewalk, a bar of wood came loose from my craft, and I fell backward, hitting and cracking a one-inch piece of marble with my head. I saw stars all right. No cosmological formula emerged, though.

I invented crude electronic devices, played with an electrostatic generator, and even designed and built a model of what I called a "Firebug" that would allow firefighters to go right into the heart of a forest fire to put it out. Since I was appalled at all the wildfires I saw ravaging the West, I sent my drawing to the National Forest Service. It was my Franklin stove—a generous present to humanity. I didn't expect any royalties, but instead I received a kind letter back from the Forest Service, passing on my creation. My inventive urge then turned to biophysics: the relationship between the living world and the global energy that infuses all of us.

The intellectual thunderstorms that drew me ever closer to Tesla's life and legacy included a desire to understand electromagnetism, global communication, physics, life energy, cosmic rays, and climate change. Although I became immersed in his story only relatively late in life, to me Tesla has become, in short order, a nexus between our current global maladies and our survival. While he didn't provide all the answers, he was certainly asking the right questions.

This book is about creative discovery seen through the lens of Tesla's life and enduring legacy. His ideas and inventions are still shaping our present and future in profound ways. There's no question that, around the globe, there's been fervent, renewed interest in the vision and work of the great inventor. In this book, I hope to illuminate his creative process and to present a pragmatic analysis that we can individually and collectively draw from as we grope for breakthroughs to problems large and small. As the creator of the operating system of the modern industrial age, Tesla deserves focused attention as an innovator and disruptor. I'm hoping to guide you into the source and evolution of his ideas in a historiography that not only looks at how he arrived at his inventions, but suggests how you can tap the same source of creation within yourself.

I will introduce you to modern-day luminaries who are investing billions in expanding upon Tesla's many powerful ideas. These intrepid souls are doers and makers, dreamers and artists. Like Tesla, they have a vision and are creating dynamic new systems for the future.

I'll revisit Tesla's relationships with giants like Edison, J.P. Morgan, George Westinghouse, Einstein, Mark Twain, Orson Welles, J. Edgar Hoover, and many others. I'll also highlight his unique bond with Chicago utilities baron Samuel Insull, who, like Tesla, was a financial failure and yet remained Tesla's friend and supporter for more than forty years. It's been a fascinating journey for me, finding traces of Tesla in Chicago, Philadelphia, the Rocky Mountains, New York City, and Belgrade, Serbia.

Ultimately, this is a journey to find the spiritual and creative core of Tesla, the tamer of flame-like power who still haunts and entreats us. It is also a crucial sojourn that we all need to take, to inform us as we struggle to survive and thrive on this verdant planet—in ever-turbulent times.

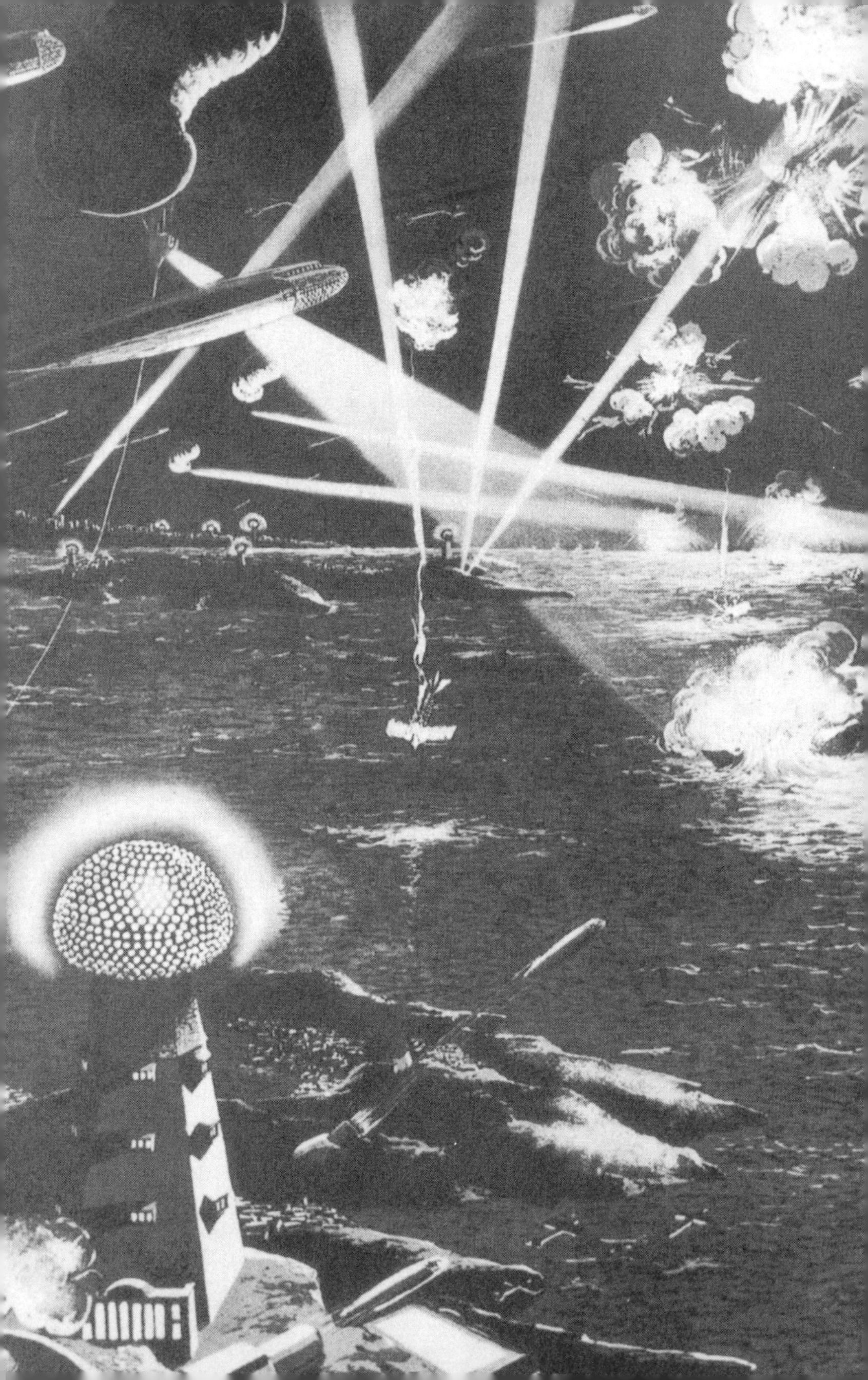

INTRODUCTION

WHY TESLA STILL ELECTRIFIES US

My search for Tesla began with a single letter and a multilayered mystery, more than a decade ago. In 2005, I was completing research for a book I was writing at Loyola University Chicago's Cudahy library archives. After looking at thousands of pages of documents, I encountered a single missive from Tesla, written in 1935, asking a failed utilities mogul for money. It struck me as a *deus ex machina*. I was fairly sure that this buried document hadn't seen the light of day for eighty years. I had never seen the letter cited *anywhere,* and its origin was obscure. But one thing was clear: I had found something akin to the Rosetta Stone in my own little world—a rubric to understanding an entirely new world yet to be revealed.

I was euphoric as I left the neo-gothic/Art Deco library on the edge of chilly Lake Michigan. It was only a few miles south on that shoreline that Tesla had triumphed in his presentation of his alternating current (AC) system in the Westinghouse exhibit at the World's Columbian Exposition in 1893. I was researching a book in which I profiled the electrical utility mogul Samuel Insull, who helped build the modern electrical grid using Tesla's AC technology. Insull had met Tesla in the mid-1880s in Thomas Edison's New York office while Edison was building his first central power station on Pearl Street in lower Manhattan. At the time, Insull had just started working as Edison's personal secretary and would eventually become responsible for managing what became the Edison General Electric Company. Once J.P. Morgan's group had consolidated and taken

This 1922 illustration by Frank R. Paul for *Science and Invention* portrays Tesla's speculative vision of the future, with towers transmitting radio-electric power for operating and controlling the sea and air defense craft, eerily presaging the advent of modern drone warfare.

Electric utility magnate Samuel Insull, shown in this 1920 photograph, was an early advocate of Tesla's AC technology.

over Edison's disorganized manufacturing businesses in the 1890s, Insull moved to Chicago, where he started a utilities empire that spanned a third of the country. It still exists today in part through the Exelon® Corporation.

When Tesla wrote Insull the letter I discovered—dated March 18, 1935—the careers of both men were like flotsam in the outgoing tide of history. Insull had been ruined by the Depression and lost control of all his companies, which were part of the biggest business bankruptcy in U.S. history at the time. Insull had experienced an epic failure, the equivalent of the Lehman Brothers bankruptcy in 2008. He was arrested in Turkey, extradited, and tried for fraud three times. He was acquitted. Insull hadn't stolen any money and had lost more than he owned.

In 1935, three years before he died penniless in the Paris Métro, Insull was flat broke and trying to make a comeback. He was reviled by everyone from Franklin Delano Roosevelt to thousands of people who had invested in the stock of his many utility holding companies, which became worthless when the Depression began. Tesla, however, still had a warm place in his heart for Insull, as evidenced in the letter I discovered at Loyola. Insull had built most of his entire empire on Tesla's AC technology, often in defiance of their old boss, Edison, long remembered in history as the genius inventor and American hero. It was Insull who created and profited mightily from an interlinked power system built around Tesla's ideas (in addition to the developments of Westinghouse Electric, General Electric, and other electrical pioneers). This system became the electrical grid: the complex network that allows us to get electricity nearly everywhere in the industrialized world.

Tesla's Intrepid Creativity

What drew me (and many others, in years past) to Tesla is his unstoppable creative energy. He was always designing, redesigning, and dreaming. Images of new inventions burst into his brain like fireworks. He lived for concepts that would provide a way of doing something better with mechanical motion and electricity, channeling the relentless energy that courses through the earth and sky. Tesla never stopped coming up with ideas, yet this 1935 letter to Insull launched my journey to find out where he was heading in his last years, how he hoped to get there, and why it matters to everyone.

With Tesla, creativity wasn't about inventions; it was about building a *system* that would disrupt the world by harnessing nature's might. He didn't just create an AC motor; he designed a network that would generate and transport electrons anywhere. Tesla also invented robotics and remote control, seeding an industry that would make drone machines work remotely halfway across the planet and (ideally) make wars less bloody. His audacity, saga, and inventions would inspire countless inventors, artists, and entrepreneurs from Elon Musk, the CEO of Tesla Motors™ and SpaceX, to Larry Page, the co-founder of Google™ and CEO of Alphabet Inc.

The Tesla Motors logo commemorates the Serbian inventor and prophet who inspired the work of today's premier technology paradigm-shifters.

And Tesla's "World System," which I will explore in later chapters, was the most audacious of all: virtually free, universal, wireless power. Tesla went far beyond lightbulbs, motors, and phonographs: He's an eternal architect of things that haven't even been created yet—machines that may allow us to tap the constantly flowing energy of the universe.

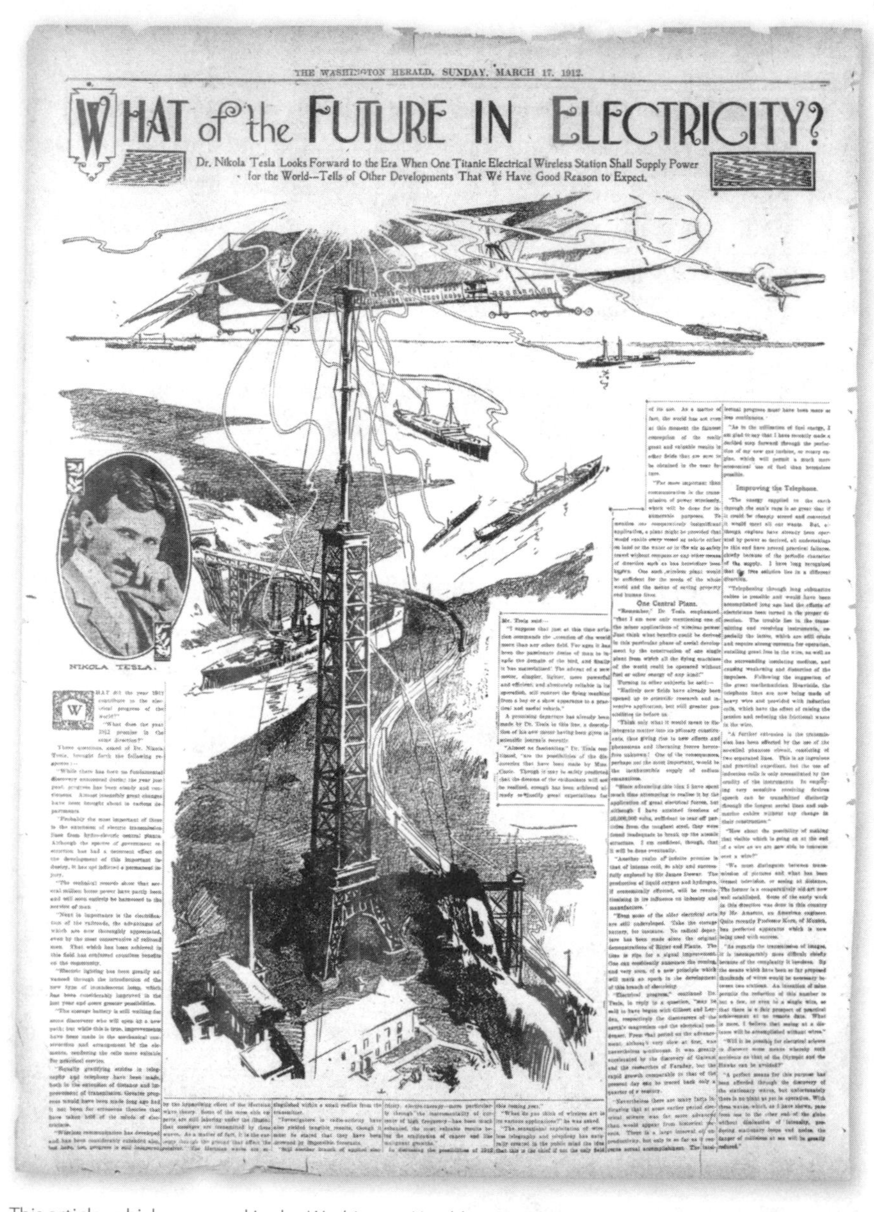

THE WASHINGTON HERALD, SUNDAY, MARCH 17, 1912.

WHAT of the FUTURE IN ELECTRICITY?

Dr. Nikola Tesla Looks Forward to the Era When One Titanic Electrical Wireless Station Shall Supply Power for the World--Tells of Other Developments That We Have Good Reason to Expect.

NIKOLA TESLA.

One Central Plant.

Improving the Telephone.

This article, which appeared in the *Washington Herald* on March 17, 1912, presents Tesla's dream of a future world where humankind is connected, united, and protected by wireless power.

World-changing inventions made Nikola Tesla a celebrity in his own time, but something otherworldly makes him transcend his era and remain a perpetual beacon for our civilization, seventy years after his death. Tesla is now an immortal rock star, an icon for billionaires, cyberpunks, artists, and "maker" inventors who are still fiddling with everyday machines in their basements and garages. Search engine designers, energy czars, musicians, and creators everywhere feel his influence. He's our Leonardo da Vinci, the Shakespeare of pure invention.

A world-class car, a rock band, and a unit of magnetic measurement have been named after Tesla. Watch any mad scientist scene in any science fiction or horror movie, and there's a good chance you'll see his Tesla coil pulsing electricity like a dynamic spider web of electrons. Tesla is oscillating energy, meters, dials, lightning bolts, and the robot-drone master. He's patron saint and mystic, discoverer and wronged entrepreneur, a bold prophet dishonored in his own time but revered in ours.

To some of his latter-day followers, it's as if Tesla never died, instead becoming some kind of a techno-mystic deity. His prescient visions and schematics of a future where energy, science, and world peace coexist elevate him above the mere title of "inventor." Indeed, few of Tesla's peers have attracted such devotion—or paranoia. New Agers insist that he talked with alien beings (or was an alien himself), while conspiracy theorists believe his idea of a "death ray" that could blast planes out of the sky was developed by the Pentagon and kept secret for nearly seventy years. Since his death, Tesla's technology has been blamed for everything from destroying Siberian forests to Hurricane Katrina.

Today, there are few stronger, sexier brands than Tesla. Year by year, Tesla's popularity grows as the memory of his contemporaries recedes even further into history. Tesla's achievements have come to overshadow those of his nemesis, Edison, who worked manically but ultimately failed to defeat Tesla's operating system. It was Tesla who laid the groundwork for the operating system of the twentieth and twenty-first centuries.

Tesla was *chimeric*—that is, he was like the ancient, mythical beast that was part lion, part goat, and part serpent. In the Greek myth, the monster is slain by the hero Bellerophon, who rides Pegasus but later falls from the winged horse. Metaphorically, to become chimeric is to embody different kinds of human creativity. *Chimeric transformation* is what I hope to explore through the life of Tesla, who endured many trials of fire as he transformed himself from an electrical engineer fine-tuning Edison's early projects to the systemic thinker dreaming up solutions for universal clean energyand world peace.

A disruptive innovator, Tesla set the tone for generations. The meta-integration of Tesla's ideas into modern technology may hold the answers to many of society's most pressing dilemmas. After seven decades, there's never been a better time to present a new profile of the groundwork laid by this stunning genius, a man whose visions seem to provide new guidance on the future of our civilization and whose astonishing ideas are still yielding world-changing innovations. Nor has there been a better time to examine his holistic creative legacy.

The mythical chimera can be seen as an embodiment of Tesla's integration of multiple creative styles.

TeslActions

In my journey to discover the creative soul of Tesla, I've scoured several archives, attended conventions, visited with the director of the Tesla Museum in Belgrade, and communed with Teslaphiles all over the world.

Tesla's tendrils reach into many places. I will show how his ideas influenced entire industries such as communications, robotics, utilities, and space travel. However, please keep in mind that this book is neither a full biography nor a technical analysis of his work—I leave that to others (see References & Abbreviations on page 245). *Lightning Strikes* is about the spirit of creativity as seen through the lens of Tesla's endeavors and monumental legacy.

To help you grasp and internalize Tesla's imaginative thought process, I will present a "TeslAction" in every chapter. These are specific ways that you can enhance and often reboot your creative process to reach a higher level of thought. Whether you're designing spaceships or just trying to solve everyday dilemmas, I hope you will find these ideas pragmatic and inspiring.

I

HOOVER'S STAKEOUT

AN ICONIC INVENTOR DIES

Before the end of this century, you will be able to communicate instantly by simple vest pocket equipment... Earthquakes will become more frequent. Temperate zones will turn frigid or torrid... and some of the awe-inspiring developments are not so far off.

—Nikola Tesla, 1926 (seventieth birthday press conference)

Government agents, who had been monitoring the Hotel New Yorker during Tesla's final decline, swarmed into his disheveled room on January 9, 1943, a day after a hotel maid discovered the lifeless body of the withered, eighty-six-year-old scientist.

Perhaps the agents had been playing cards in the vast bowels of the technologically advanced hotel, which featured its own power plant. Maybe they were reading the paper. Their orders came straight from the Office of Alien Property (OAP), an obscure wartime agency set up to screen and monitor possible foreign agents in the United States: *Take all of his papers and any models of inventions.*

The agents probably found it strange that they'd been asked to descend on the old inventor's last address for what amounted to a raid. After all, Tesla had tried to work *with* the government to develop weapons for the American military. Though he'd been born in the Balkans, Tesla, long an

Tesla's final home: the splendid, iconic Hotel New Yorker, 1930.

FBI director J. Edgar Hoover, 1940.

American citizen, had some familial connections to now Nazi-occupied lands. Yet he was hardly a communist or a fascist; in fact, his politics resembled patriotic pacifism. In the last three decades of his life, his ideas centered on protecting countries from invasion, although his inventions would later launch a new era of robotic war.

Before his decline, which began after funding was cut off for his "World System" (see chapters VI and VII), Tesla had energized the world with his AC system. He came up with the basic technology for radio, wireless power, robotics, and dozens of smaller devices. He even experimented with X-rays and radio telescopes and designed the plant that converted the hydrodynamic energy of Niagara Falls into electricity that could be transported hundreds of miles. Tesla was the genius behind the primary engine of the Second Industrial Revolution.

The ever-paranoid FBI director J. Edgar Hoover was taking no chances in the middle of a war. Hoover knew from his ongoing surveillance of the inventor's last years that Tesla had associations with fascist elements in the United States—friends who had openly praised Hitler. What if some of Tesla's weapons inventions actually worked and German agents got hold of his plans? Hoover didn't want to be responsible for that blunder, especially since the Nazis were ahead of the United States in rocket research. Whatever Tesla had, Hoover didn't want the enemy to acquire it.

Obsessed with the number three, Tesla had selected his room in the modern, towering hotel specifically for its number: 3327. Three was important on its own, but numbers divisible by three (like 27 and even 3,327) were *especially* mystical—some kind of Platonic ideal, a holy trinity of numbers. But the agents, to whom Tesla was probably nothing more than an eccentric old man, were unconcerned with the room number's supposed mystical properties and performed their tasks dispassionately. They would have been put off by the scattered pigeon feed—treats for Tesla's last true companions—as they gathered up whatever secrets the inventor had stored there. According to Tesla biographer Marc Seifer, "Tesla had some documents in his tiny hotel rooms, but 80 trunks were already at a Manhattan warehouse. The rent was being paid by Tesla's nephew Sava Kosanovich, who was the ambassador to the United States from Yugoslavia."

The OAP, which had expertise in impounding Europeans' property in the United States, would hold his hundreds of notebooks for further evaluation for more than a decade. If there was any evidence that Tesla was connected to a spy network, as Hoover feared, the government could shut it down quietly.

What was in the papers? Tesla was well known for his announcement of what became known as his "death ray," which he had extolled to the press on his eighty-fourth birthday. He had described the device, vaguely, as a gift to the world that was so horrifying that its presence would guarantee peace.

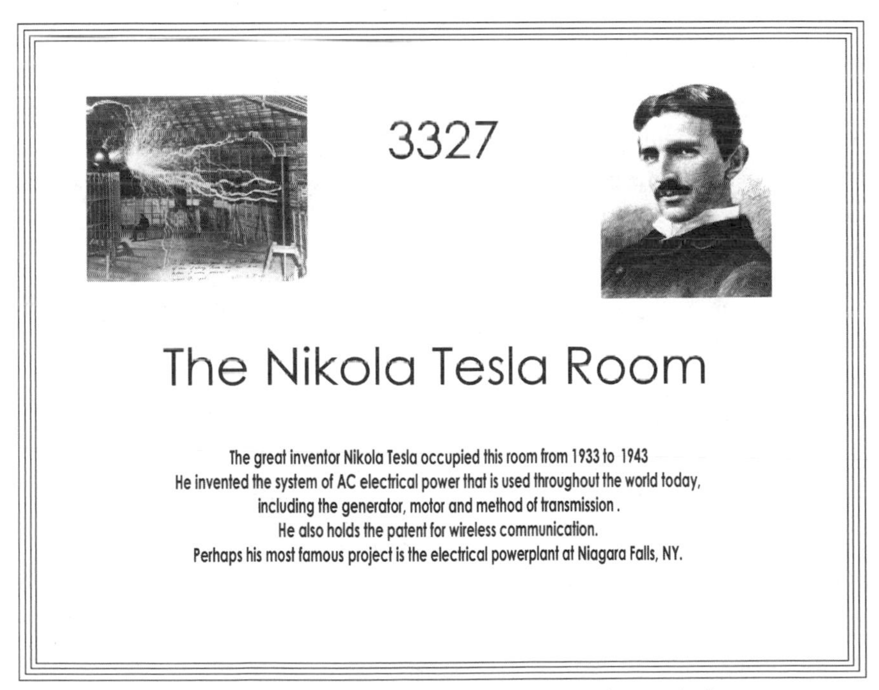

This memorial plaque adorns the door to the hotel room in which Tesla spent his last years.

"The death beam is based entirely on a principle of physics that no one has ever dreamed about," Tesla told the *New York Times* three years before he died. Few details were forthcoming, although Tesla assured reporters that a plant could be built for $2 million to create the energy for the weapon. A dozen of these plants at strategic points across the country could defend the continent from attack. But no one had ever seen a working model.

And then there was telegeodynamics, Tesla's grand plan to send energy through the earth itself to any point on the planet or even "split it like an apple." Could Hitler use this technology, if it was viable at all, to create earthquakes to defeat the Allies or generate a tsunami before he invaded North America?

Almost certainly, Tesla had written plans—notes, schematics, and detailed drawings. Unlike Leonardo da Vinci, Tesla didn't make these explorations into the relationship between nature and energy out of pure curiosity or aesthetic refinement; he wanted to know the ecology of the constant stream of power that lit up the universe for a purpose.

After decades of research into Tesla's "death ray" concept by dozens of experts, it seems that much of Tesla's vision was in his head. There were some rough drawings and specifications for a huge amount of power, but no one quite knew how this weapons system would actually work. If he had notebooks delineating his master systems, they may not have contained exquisite drawings like Leonardo's, but rather the formulas and electronic diagrams of a more scientifically and technologically advanced era.

Tesla boasted of his new ideas at birthday celebrations that he staged as media events in the 1930s. The man's mind never stopped. He rarely slept more than a few hours a night, if he slept at all. Tesla didn't think of one invention at a time like Edison, and he wasn't interested in peddling the next recording or illumination device; he envisioned all-embracing concepts. He wanted to broadcast power and information around the world with one system—an idea that would disrupt everything that the New York financial cartel had backed.

The OAP men went about their methodical removal of Tesla's notes despite ample evidence that Tesla's eccentricities might finally have gotten the better of him. Tesla was a man who insisted on buying a new necktie every week (always for one dollar), who had once been tortured by the ticking of a clock, and who refused to eat until twenty-four napkins had been set to his left on the table. Tesla never ate at all in the company of women wearing pearls, which repulsed him. His eccentricities aside, the bosses in Washington were certain that Tesla had the know-how to power some awesome weapons—possibly even incredible devices that could shoot planes out of the sky from a distance of hundreds of miles.

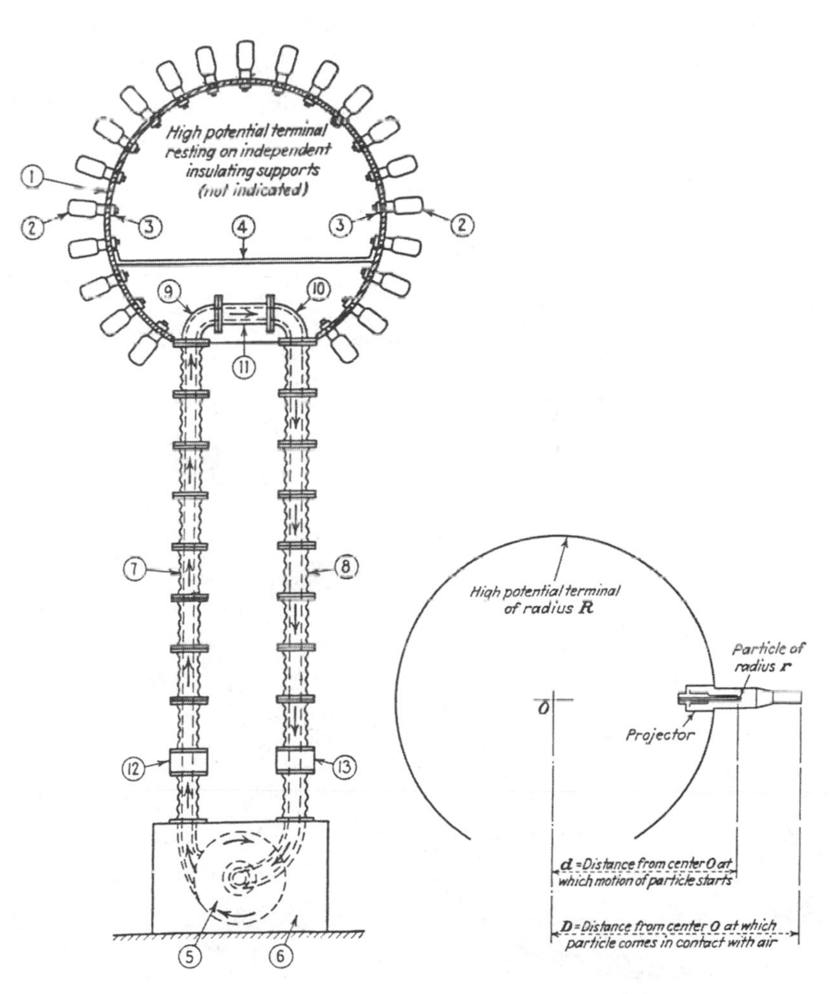

A schematic illustration from 1937 showing Tesla's plan for a high potential electrostatic generator to be used as part of his proposed particle beam "superweapon."

Was Tesla Mentally Ill?

Was there an element of mental illness that was a moving part in Tesla's stunning creativity? When studying Tesla, it's far too easy to put him into a box of psychiatric maladies found in the *Diagnostic and Statistical Manual of Mental Disorders* (DSM). It's easy to make a superficial judgment on Tesla's mental state. Was he bipolar or obsessive-compulsive (or neither)?

Tesla worked maniacally and suffered occasional breakdowns. He had an aversion to pearls and didn't shake hands. Multiples of threes were important to him. According to biographer W. Bernard Carlson, these obsessions germinated as he struggled to make up for the loss of his brilliant and beloved brother, Dane, who died in a riding accident at age twelve when Tesla was just seven years old. It was Dane, deemed "extraordinarily gifted," who was to follow his father into the priesthood.

Tesla working diligently in his tidy office at 8 West 40th Street, 1916. The inventor exhibited signs of OCD, especially late in life, but was not handicapped by it.

> Tesla sought to win back the love of his parents by striving to be perfect.... Unable to please his father, Tesla "contracted"...what now might be called obsessions...[that] undoubtedly interfered with his relationships with other people.

As Tesla himself recalled,

> "The recollection of [Dane's] attainments made every effort of mine seem dull in comparison. Anything I did that was creditable merely caused my parents to feel their loss more keenly. So I grew up with little confidence in myself."

Clearly, the death of his older brother was a defining moment that haunted Tesla for the rest of his life and activated within him a perfectionistic streak, which manifested itself in various obsessions. The favorite, eldest son, Dane had been handsome, witty, and intellectually precocious—a hard act to follow.

Tesla also confessed to experiencing the extremes of emotions, beginning when he was a young boy:

> "My feelings came in waves and surges and vibrated unceasingly between extremes. My wishes were of consuming force and like the heads of the hydra, they multiplied..."

It is hardly unusual for highly creative individuals to struggle with bipolar disorder, as Tesla noted in an interview in 1896, responding to a question about the frequency of his depressive bouts: "Every man of artistic temperament has relapsed from the great enthusiasms that buoy him up and sweep him forward."

Despite his unusual obsessions, perfectionism, and emotional peaks and troughs, Tesla was still able to function at a very high level, recovering from his breakdowns to attain even greater creative heights.

To understand his mental machinery, it is necessary to examine much of what is perceived to be abnormal about him in context. The visions he had didn't lead to any psychotic behavior. Voices didn't tell him to do unspeakable things. Indeed, the vast majority of people with mental illness are not harmful, nor are they particularly inventive, although this link has always persisted in our culture.

The Serbian clinical psychologist Dr. Milena Tatic Bajich is part of the greater Tesla community, having founded a Tesla Club in Chicago. She claims that Tesla probably fits the profile ascribed to obsessive-compulsive personality disorder (OCPD). "Obsessive-compulsive disorder (OCD) prevents people from functioning on a daily basis," Dr. Bajich told me. "Sure, Tesla had some quirks, but let's cut him some slack. OCD is an acute form and can be chronic in nature. OCPD is a condition that explains the character of the person: rigid, perfectionistic, micro-managers." In other words, OCD is a crippling disorder that can prevent an individual from functioning normally, while OCPD describes dominant character traits that are not necessarily dysfunctional. As with all definitions in the world of psychology, however, the boundaries are often fuzzy.

People with actual OCD are hostage to their repetitive loops. They can't function in the world as most people do because they are shackled by repetitive behaviors. Certainly, Tesla was a workaholic who drove himself to exhaustion, and he hung onto big ideas when others had abandoned him. He was tenacious and obsessive. He was able to work insanely long hours with little or no sleep and could conceive of a machine in both a visual and mathematical sense. But that's not a person suffering from OCD, Bajich maintains. Eccentric? Yes. Mentally disturbed? No. In fact, his own reflections on his odd habits and obsessions suggest a remarkable degree of self-awareness that was further proof of his lucidity and self-control.

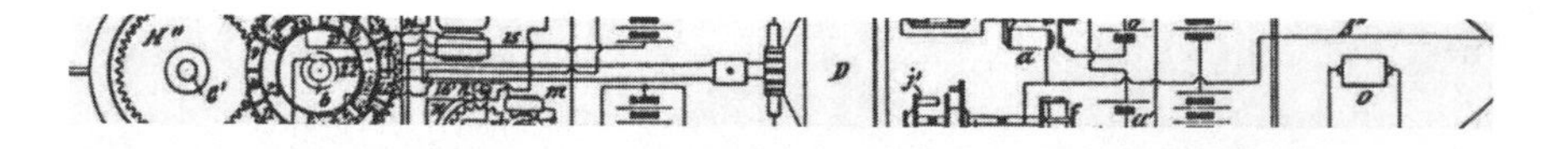

IN HIS OWN WORDS: ECCENTRIC TESLA

In his 1919 autobiography *My Inventions*, Tesla reflected at length on his own psychology, as well as his childhood experiences and how they shaped him. He also made a number of observations about his parents. As Tesla recalled, his father "had the odd habit of talking to himself and would often carry on an animated conversation and indulge in heated argument, changing the tone of his voice."

Following the death of his brother, Tesla recalls developing some very strange associations that caused him acute discomfort:

> **"During that period I contracted many strange likes, dislikes and habits, some of which I can trace to external impressions while others are unaccountable. I had a violent aversion against the earrings of women but other ornaments, as bracelets, pleased me more or less according to design. The sight of a pearl would almost give me a fit but I was fascinated with the glitter of crystals or objects with sharp edges and plane surfaces. I would not touch the hair of other people except, perhaps, at the point of a revolver. I would get a fever by looking at a peach and if a piece of camphor was anywhere in the house it caused me the keenest discomfort. Even now I am not insensible to some of these upsetting impulses."**

Tesla's ability to mix sensory experiences, as when someone with synesthesia listening to music involuntarily associates the various notes with different colors or smells, was an integral part of his sensory experience. Reciting poetry could bring visions into his mind, for example. However, Tesla was able to use his gifts to great advantage, even if some critics wanted to wrongly classify them as disabilities (a controversial label on its own). When you are dealing with the complex world of electromagnetism, this unique trait may actually be a distinct benefit.

As a boy, Tesla was lucky enough to have access to his father Milutin's personal library, though he was punished when Milutin, a devout Serbian Orthodox priest, caught him reading any of the books. Punishment did not deter him, however. He had an enormous appetite for knowledge and an obsessive need to complete everything he started:

> **"I had a veritable mania for finishing whatever I began, which often got me into difficulties. On one occasion I started to read the works of Voltaire when I learned, to my dismay, that there were close on one hundred large volumes in small print."**

Perhaps hoping to quell his mania, and determined to get out from under his dead brother's very long shadow, Tesla encountered an inspirational work of fiction that focused on a young man who overcame debauchery and love of pleasure, lifting himself to national hero status through sheer strength of will and inflexibility of purpose. This story would guide him toward an ascetic lifestyle and guiding philosophy, as reflected in the tireless work ethic and rigid self-control for which he was well known:

> **"On one occasion I came across a novel entitled *Abafi* (the Son of Aba), a Serbian translation of a well known Hungarian writer, Josika. This work somehow awakened my dormant powers of will and I began to practice self-control."**

Although Tesla's intense preoccupation with self-control led to certain compulsive behaviors and interfered in his relationships with others, this trait was one of the keys to his success. In the late 1960s and early 1970s, Stanford psychology professor Walter Mischel conducted an infamous study in which children were given the choice between getting a small reward (such as a marshmallow) immediately and a larger reward (e.g., two marshmallows) after fifteen minutes, and then followed up with participants years later to evaluate their life outcomes. Those who waited, or delayed their gratification, had better life outcomes than those who quickly chomped on the marshmallow. The study suggested a definite correlation between self-control and achievement. As Telsa put it, "to do our best work...we must exercise moderation and control our appetites and inclinations in every direction."

In addition to being highly disciplined and curious, young Nikola was bashful. He developed a close relationship with the family cat, Macak, that foreshadowed his unusual tenderness for the pigeons that gathered at Bryant Park near the end of his life.

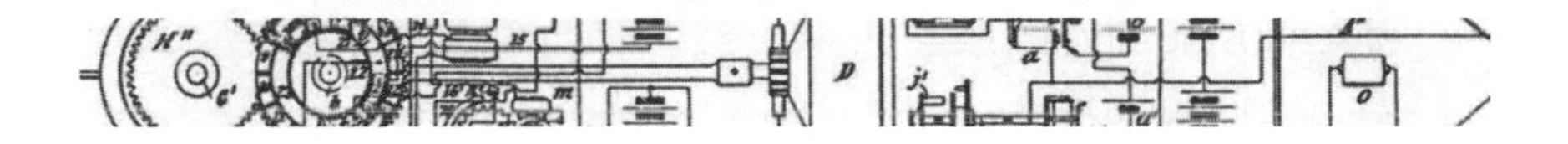

A Mystery Unfolds

The FBI confiscation and lost paper trail was the second leg in my journey to unearth the essence of Tesla. After I had discovered the 1935 Insull letter, I was convinced that there was more to the story. Like many before me, I sought FBI records through a Freedom of Information Act (FOIA) request and received a pound of letters, clippings, and ephemera from the secretive agency in 2009. Many of the documents were redacted—governmentese for the blacking out with a marker of certain names by some bureaucrat under the guise of national security.

Why would the government care to cloak the identities of those involved in Tesla's papers seventy years after they were confiscated? Of course, dozens of conspiracy theories abound: *Tesla was murdered, and the government wanted to cover it up. He really did have plans for a powerful, destructive weapon. Hoover discovered that he had something of value and deep-sixed the documents in a government warehouse, never to be found. The names of FBI agents of the time were blotted out because they were into other, more questionable spying.* Hoover, after all, regularly spied on *presidents*, mostly to blackmail them into hewing to his anticommunist agenda.

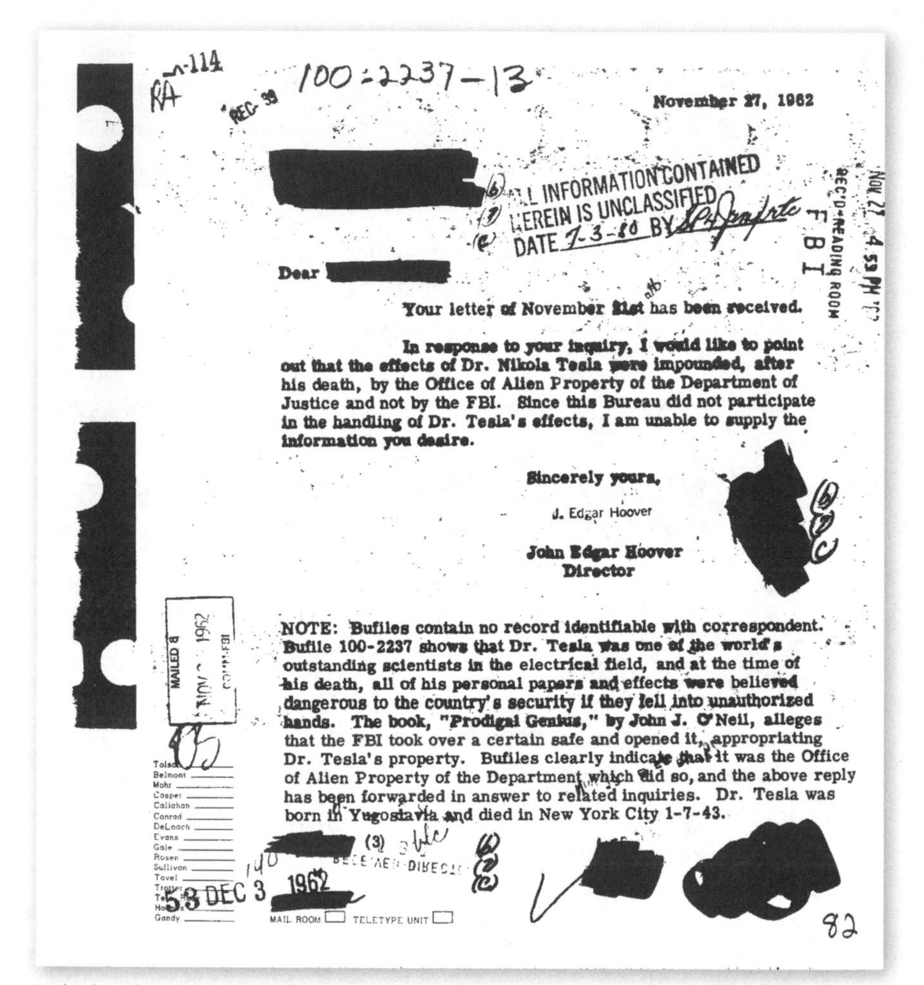

100-2237-13

November 27, 1962

ALL INFORMATION CONTAINED HEREIN IS UNCLASSIFIED DATE 7-3-80 BY

Dear

Your letter of November 21st has been received.

In response to your inquiry, I would like to point out that the effects of Dr. Nikola Tesla were impounded, after his death, by the Office of Alien Property of the Department of Justice and not by the FBI. Since this Bureau did not participate in the handling of Dr. Tesla's effects, I am unable to supply the information you desire.

Sincerely yours,

J. Edgar Hoover

John Edgar Hoover
Director

NOTE: Bufiles contain no record identifiable with correspondent. Bufile 100-2237 shows that Dr. Tesla was one of the world's outstanding scientists in the electrical field, and at the time of his death, all of his personal papers and effects were believed dangerous to the country's security if they fell into unauthorized hands. The book, "Prodigal Genius," by John J. O'Neil, alleges that the FBI took over a certain safe and opened it, appropriating Dr. Tesla's property. Bufiles clearly indicate that it was the Office of Alien Property of the Department which did so, and the above reply has been forwarded in answer to related inquiries. Dr. Tesla was born in Yugoslavia and died in New York City 1-7-43.

MAILED 8
NOV 1962

Tolson, Belmont, Mohr, Casper, Callahan, Conrad, DeLoach, Evans, Gale, Rosen, Sullivan, Tavel, Trotter, Gandy

53 DEC 3 1962

MAIL ROOM TELETYPE UNIT

82

In this heavily redacted letter from 1962, Hoover claims to have no knowledge of the whereabouts of any of Tesla's missing papers or artifacts.

During the Cold War, there was even more interest in the Tesla files as Hoover fought to keep them out of the hands of possible Soviet agents, who may have already gotten access to the documents. Each request for access to the FBI file from 1943 to the present carried the same response: The agency didn't have the documents and wouldn't say where they went after they were shipped to other government warehouses.

Undaunted, I sent FOIA requests to the National Archives, Defense Advanced Research Projects Agency (DARPA), and various branches and bases of the Department of Defense. DARPA, the agency behind the creation of the Internet, said it didn't have anything, as did the National Archives, which may have inherited some of the papers of the Office of Alien Property. I have yet to hear from the Pentagon, despite repeated requests and an effort to enlist my congressman to track down the records.

According to Seifer, who wrote his PhD thesis on Tesla and several subsequent insightful articles and books on the inventor, several people had evaluated Tesla's papers shortly after the papers were taken in 1943, including John Trump, an MIT engineer, and Bloyce Fitzgerald, an army engineer. Although other researchers have said that Tesla probably experimented with particle beams in the late nineteenth century, they couldn't verify with any certainty if he had fully designed a working weapon based on his findings, even though Tesla said on his eighty-first birthday in New York that he intended to bring his weapon "to a Geneva conference for world peace." The inventor had described the weapon more than once in his writing in the 1930s.

In one of the badly reproduced, undated, and redacted FBI memos I obtained, Fitzgerald said he "knows that Tesla has conceived and [blacked out word] a revolutionary type of torpedo which is not presently in use." Was this weapon a robotic torpedo? How was it powered? The FBI memo doesn't say, nor does it explain what the FBI hoped to find out or how it followed up.

What has fueled the interest in Tesla's lost archives was the Soviet Union's interest in particle beam weapons in the 1970s and the "Star Wars" Strategic Defense Initiative, conceived by the Reagan administration to shoot down intercontinental ballistic missiles with energy beams. A 7,000-word piece from 1977 in the respected trade magazine *Aviation Week and Space Technology* highlighted the Soviet effort at the time. The navy even recently announced that it had such a weapon that could be deployed on new destroyer warships, although it's mum on whether Tesla's ideas had anything to do with its development. We may never know.

The ever-debonair Tesla with his cane at the Hotel New Yorker during the 1930s.

The Other Tesla That Lives On

Tesla's name is on a car. He's a recurring character in movies and television shows. Special-interest groups from cyberpunks to avant-garde rocketeers revere him. He's so cool, he's hot—all over the world. And there's a reason: If you investigate Tesla's most creative periods, you discover a man who wanted to solve some huge problems that still vex humanity. How do we harness and channel a ubiquitous source of clean energy? How do we avoid wars? How do we improve our overall quality of life? Compelling answers are hinted at in Tesla's copious notes and diagrams.

Still, the image of Tesla's final days and financial collapse tends to hurt his brand and ultimate place in history. As a recluse in his final decades, Tesla had retreated into a world free of reluctant financiers, jealous inventors, thieving charlatans, and spectacle worshippers. Although the inventor had a high-profile funeral in New York's Cathedral of Saint John the Divine, he was nowhere near as celebrated as his contemporaries Edison, Morgan, and Twain. Like the cathedral that was under construction as he lay in state, Tesla's legacy is still unfinished business.

And then there were the pigeons. Even as his health faded and he scraped by on a few royalty dollars, Tesla insisted that his assistants feed the fowl that flew into the window of his hotel apartment, and he made midnight treks to Bryant Park to attend to his winged sycophants. The G-men surely couldn't avoid evidence of this infatuation as they packed up his belongings—his window ledge was caked with pigeon guano. Tesla had even come to think of one of the adopted pigeons as his wife. "I loved her as a man loved a woman," Tesla told his confidant John O'Neill:

> "When she was ill, I knew and understood; she came into my room and I stayed beside her for days. I nursed her back to health. If she needed me, nothing else mattered. As long as I had her, there was a purpose in my life. Then one night as I was lying in bed in the dark, solving problems, as usual, I knew she wanted me; she wanted to tell me something important so I got up and went to her. As I looked at her, I knew what she wanted to tell me–she was dying. And then, as I

> got her message, there came a light from her eyes–powerful beams of light. It was a real light, a powerful, dazzling, blinding light, a light more intense than I had ever produced by the most powerful lamps in my laboratory. Something went out of my life. Up to that time I knew that I would complete my work, no matter how ambitious the program, but when that something went out of my life, I knew my life's work was finished. ❞

The agents might not have been old enough to know, however, that this lonely, withered figure had once been a glamorous giant in New York society as well as on the world stage. Women had adored the handsome European who spoke six languages, yet remained celibate—or so he claimed—in order to focus fully on his life's work. He'd been slim and striking, standing more than six feet tall, with large hands, abnormally large thumbs, and a peculiar wedge-shaped head.

In better days, Tesla lived at the Waldorf Astoria Hotel in high style, coming down to dine at his own table every evening at precisely eight o'clock, dressed in a formal tailcoat. His laboratory feats entertained his many famous friends, including Mark Twain, who was fascinated by new inventions of the second industrial age—and lost a bundle investing in them.

Tesla had been a mystic as well as an inventor, and that was part of his allure. He believed in a higher power and sought a way to achieve world peace. Studying the work of Swami Vivekananda, he read the Hindu Vedas and wrote poetry. Unlike his rival Edison, he dabbled in cosmology, religion, and the mind-body connection.

Tesla's favorite white pigeon.

But the Tesla the agents knew best was the haunted figure who had just died in Room 3327. In the years preceding his death, Tesla would wander from his hotel room like a specter and stroll the streets of New

A man feeds the pigeons at Washington Square Park, 1941.

York deep into the night. When he died, Tesla's distinctive, noble face was hollowed out and emaciated. He had long since disconnected himself from society, and many viewed him as a bitter scientist who had finally lost touch with reality. Even as most of the outside world regarded him as a blemished mystic, Tesla kept the prophecies coming. But, as death approached, he shared the exquisite details of the visions that kept him awake most nights only with his notebooks and his birds.

TeslAction 1 Be Introspective

In 1919, Tesla, reflecting on his life, made this astute observation:

> “From childhood I was compelled to concentrate attention upon myself. This caused me much suffering but, to my present view, it was a blessing in disguise for it has taught me to appreciate the inestimable value of introspection in the preservation of life, as well as a means of achievement.”

We are all different, and the most highly creative people often grow up feeling a little bit alienated. They may have been teased for being "weird" or "quirky." They either detest conformity or find it very difficult to conform to others' expectations. Tesla had many strange obsessions and a mania for knowledge, and he seemed to prefer the company of his cat over that of other people, but he was able to leverage these unusual predilections to become one of the most iconic inventors who ever lived.

Even if you didn't have purple hair or a reputation in the classroom for getting lost in your own private world, chances are you had some unique experiences that shaped you as a person and gave you a unique perspective on things. In fact, these unique qualities and perspectives can be extremely useful when we have to come up with creative ideas and solutions to problems. For Tesla, as noted earlier, a major turning point that shaped him as a person came after his talented brother, Dane, died in an accident. In his effort to compensate for his brother's absence, Tesla, suddenly becoming his parents' only son, developed a relentless work ethic and a hunger for recognition. He also became acutely aware of his own mortality and made it his mission to make the world a better (and safer) place for all humanity before his own time ran out.

Take a moment to reflect on who you are as a person and how you came to be that way. Take stock of how you differ from others and how you can use that distinctiveness to your advantage. Here are some questions to guide you:

- What are your core qualities or dominant personality traits? Write down as many as you can think of, and then circle three of them that best encapsulate who you are as an individual.
- What are some of your most memorable and unusual life experiences? Can you link any of these to your unique personality, perspective, or interests?
- What are the ways in which you most enjoy spending your free time? What are your favorite hobbies?
- Do you have any habits or tendencies that others might perceive as "weird"? What do these eccentricities suggest about you as a person?
- Look back at your list of personal qualities. What kinds of endeavors might these be most suited for? Where can you contribute the most value?
- What propels you into a state of intellectual or spiritual reverie? How does it relate to meaning and purpose in your life?
- Where do you see yourself in ten years? What kind of impact on the world would you like to have had?

II

CATCHING FIRE

TESLA'S CREATIVE INHERITANCE

Shakespeare's hero Prospero is not just an abstract character. There have always been those kind of people, but Nikola Tesla is one of the most illustrative examples of someone with a skill set, ability and imagination comparable to Prospero's.

Zorica Civric, Senior Curator,
Museum of Science and Technology, Belgrade

Tesla had many forebears who also applied their curiosity, creative vision, and experimental rigor to numerous natural phenomena. The inventor was inspired by those ancient and eminently practical Arab sailors who calculated their voyages by the motions of the stars, and those Roman engineers who designed sophisticated methods for moving fresh water to where it was needed and siphoning off sewage from their cities.

Although it's not known if Tesla ever saw the Renaissance-era Italian's notebooks, Leonardo da Vinci was a kindred spirit. Leonardo's interests were diverse and scintillating. Nothing was off-limits to da Vinci, who, like Tesla, wanted to divine the secrets of nature. The polymath was fascinated with natural processes: water coursing, flowers unfolding from seeds, the structure of the human skull. He drew natural and mechanical items detailed with an artist's eye and a scientist's curiosity, writing strangely in his backward

Benjamin Franklin Drawing Electricity from the Sky by Benjamin West, c. 1816.

script. The artist of *Mona Lisa* and *The Last Supper* probed anatomy, botany, hydrodynamics, mechanical engineering, and zoology. How did they all fit together, Leonardo wondered. How did these systems work? Waterwheels, levers, gears, and pulleys dominated his notebooks, which were adjunct studies for his war machines for the Sforza family, which dominated Milan in the late fifteenth century. Collected in what is now called the Codex Atlanticus, the notebooks (one of several collections) span some forty years of Leonardo's life—from 1478 (when he was twenty-six) to his death in 1519.

From an early age in my life, Leonardo was one of my heroes, and in 2007 I had a chance to almost touch his drawings when some of the Codex was put on display in the Maritime Museum in Barcelona. My wife and I were visiting Spain as part of a twentieth-anniversary trip when we stumbled upon the exhibit serendipitously. What fascinated me most about Leonardo's sketches was the desire to know the ecology of what he was looking at. What gave moving water its power? How did birds fly? How could a *man* fly? When he was supposed to be delivering works of art for his well-heeled clients—he frequently missed his promised delivery dates—he was puttering around with the wheelworks of nature. His anatomical drawings explore nearly every tendon, every muscle. The fibers that made up everything from the human hand to a milkweed seed entranced him. Leonardo also made impeccable drawings of devices that could move water—or anything else, for that matter—in any direction via water screws, wheels, and canals. His patron, the Sforza family, certainly had need of such engineering to modernize greater Milan.

Likewise, Tesla was fascinated by the mysteries of nature, as well as the scientific efforts to harness it. As discussed in chapter I, from an early age he had a voracious appetite for knowledge. This appetite extended beyond reading the great works of science and literature to actual tinkering. As he recalled in *My Inventions*,

> “I undertook to take apart and assemble the clocks of my grandfather. In the former operation I was always successful but often failed in the latter.... Shortly thereafter I went into the manufacture of a kind of pop-gun which comprised a hollow tube, a piston, and two plugs of hemp.”

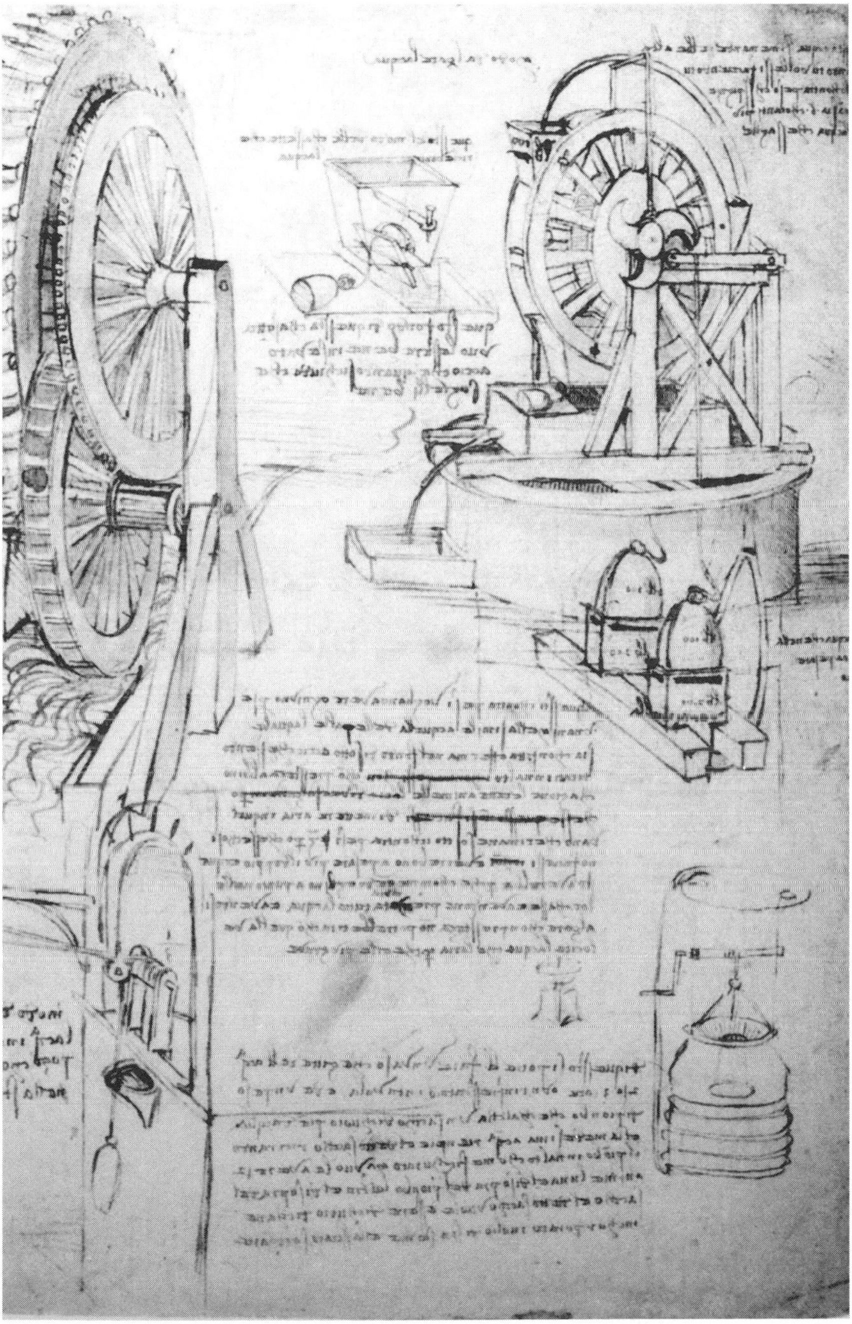

Hydraulic devices for transporting water by Leonardo da Vinci, 1480.

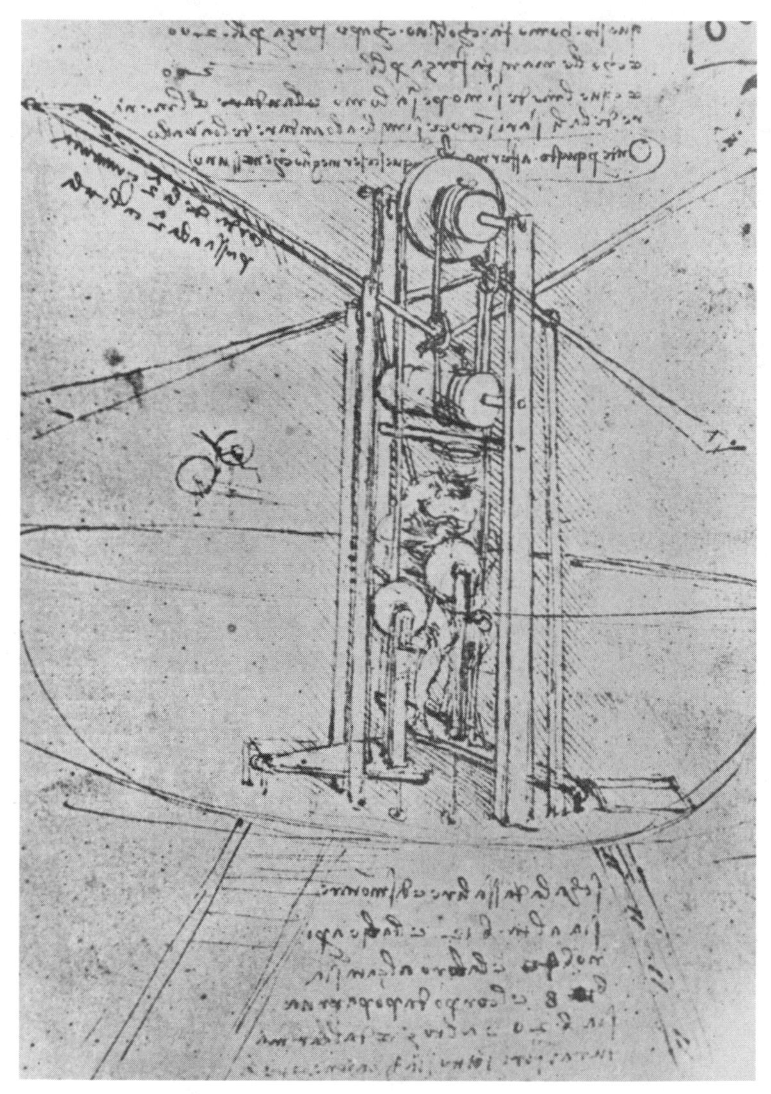

Da Vinci's sketch of a flying machine, c. 1487.

Propelled by his overly active imagination, Tesla applied himself to creating flying machines, as da Vinci had famously done:

> Mechanical flight was the one thing I wanted to accomplish although still under the discouraging recollection of a bad fall I sustained by

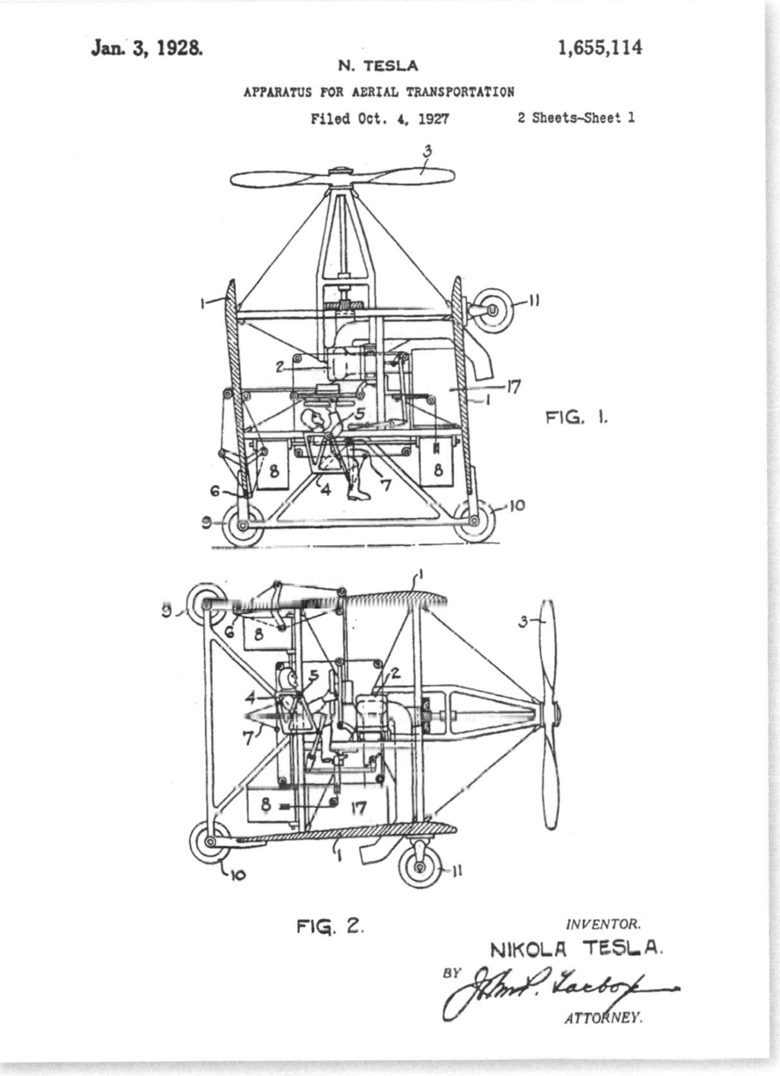
Jan. 3, 1928.

1,655,114

N. TESLA

APPARATUS FOR AERIAL TRANSPORTATION

Filed Oct. 4, 1927

2 Sheets-Sheet 1

Tesla filed this patent for his flying machine in 1927.

jumping with an umbrella from the top of a building. Every day I used to transport myself thru the air to distant regions but could not understand just how I managed to do it. Now I had something concrete—a flying machine with nothing more than a rotating shaft, flapping wings, and—a vacuum of unlimited power! ”

Like da Vinci, he also studied the flow of water:

> “In the schoolroom there were a few mechanical models which interested me and turned my attention to water turbines. I constructed many of these and found great pleasure in operating them.... I was fascinated by a description of Niagara Falls I had perused, and pictured in my imagination a big wheel run by the Falls.”

Later in life, Tesla would make this wheel a reality, harnessing the tremendous force of the mighty Niagara Falls to transport electricity far and wide.

Leonardo, like Tesla, was one of the first integrated systems thinkers. He wanted to know the architecture of animate and inanimate things and hoped to discover the rules governing these systems. He wanted to understand not only how mechanical things worked, but also how to make them more efficient and part of a larger working whole. Physicist Fritjof Capra observes:

> Our modern systemic conception of life fully validates Leonardo's method of exploring similarities between patterns and processes in different living systems, and his view of the earth as being alive has reappeared in today's science, where it is known as Gaia Theory.

Gaia, indeed. Nature is certainly a self-regulating system, although humans are adept at messing with Mother Earth's gearbox. We expel carbon dioxide from our lungs and from the process of combustion. Plants put oxygen into the air and digest nitrogen, which dominates the atmosphere. Energy is hitting us all the time from man-made sources and the sun. Earth hums with energy and has a massive magnetic field that protects us from deadly solar radiation endlessly bombarding us. Tesla understood some of this, and like Leonardo, he wanted to know how all the moving parts worked in harmony. Nature, to Tesla, was a system of relationships, an ecology he yearned to understand. He recalls a "eureka moment" that struck him one day while he was roaming in the mountains of Croatia:

> “The sky became overhung with heavy clouds but somehow the rain was delayed until, all of a sudden, there was a lightning flash and a few moments after a deluge. This observation set me thinking. It was manifest that the two phenomena were closely related, as cause and effect, and a little reflection led me to the conclusion that the electrical energy involved in the precipitation of the water was inconsiderable, the function of lightning being much like that of a sensitive trigger.”

This natural cause and effect led Tesla to wonder if humans could use electricity to trigger precipitation over arid lands, in essence bending the elements to the will of humanity.

Tesla's sense that humans could exercise powerful control over nature was reinforced by another event from childhood. He and some friends were playing in the snow, each trying to create a bigger snowball than the next person by throwing balls of snow down a steep mountainside. One of these snowballs, according to Tesla, gathered so much snow on the way down that it became as big as a house:

> “For weeks afterward the picture of the avalanche was before my eyes and I wondered how anything so small could grow to such an immense size.... had it not been for that early powerful impression, I might not have followed up the little spark I obtained with my coil and never developed my best invention.”

Like many scientists before him, Tesla drew much wisdom from his observations of nature. Witnessing the power of snow and gravity sparked a desire to understand how a colossal magnification of "feeble" forces is accomplished, and he devoted much of his life to that effort.

By the end of the century in which Leonardo died, Shakespeare would be musing on the "brave new world" in which magic was becoming more than the conjurer's and dramatist's art. Science was overtaking alchemy as a legitimate and powerful source of human inquiry and knowledge. In the bard's last play, *The Tempest*, his exiled nobleman Prospero literally conjures a storm, using magical powers to change the destiny of those on board a ship headed near his island. Tesla later claimed that such "magic" was possible with his science. *Control the weather; make nature bend to your will.*

Shakespeare seized upon a theme that would dominate the Scientific Revolution to come after he died: how to understand and manipulate nature. As critic Harold Bloom notes, "The art of Prospero controls nature, at least in the outward sense. Though his art ought to teach Prospero an absolute sense of self control, he clearly has not attained this as the play concludes."

On the dynamic stage of the late nineteenth and early twentieth century, Tesla is a modern Prospero—a magus who should be bitter, bent on manipulating nature because he's been maligned. Yet Tesla is the stage manager of self-control and discipline. He wards off disturbing images as a young man, overcomes binge gambling and frequent nervous breakdowns, practices celibacy, and becomes a vegetarian. Like Prospero at the end of *The Tempest*, Tesla is forgiving. He wants peace, although all of his discipline and vigorous exploration doesn't leave him with that dramatic serenity.

Prospero (right) summons the storm and shipwreck at the beginning of Shakespeare's *The Tempest* in this 1797 engraving.

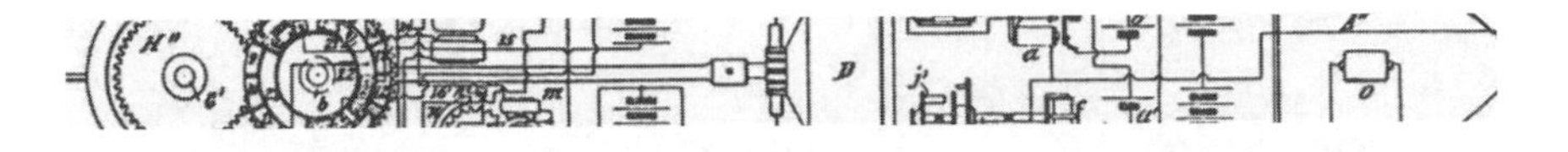

BEN FRANKLIN'S KITE

Sometimes pure curiosity will take you a long way in the realm of self-disciplined scientist, even if you don't have the artistic sensibilities of Leonardo. When Benjamin Franklin started his experiments in electricity, he just wanted to understand the nature of lightning. What kind of cosmic force was it? Was civilization doomed forever to be smote by lightning bolts, an unbridled and uncontrollable force that could cause fires and level entire city blocks at random? Franklin didn't believe so.

Franklin's groundbreaking electrical experiments in the mid-1700s, at a time when America was beginning to chafe under British rule and the Revolution wasn't yet a conflagration, proved that electricity was a definable force of nature. Surprisingly, these experiments were not described in some formal treatise or technical journal, but rather in personal letters to a London friend, Peter Collinson, starting in 1747.

It's clear from these letters that Franklin spent hundreds of hours trying to see how electricity was stored, transported, and transmuted. Although he focused on static charges—he introduced to the world the concepts of positive and negative polarity—he laid the groundwork for the conviction that you could *do* something with this force of nature. It could be harnessed for other uses.

To Franklin, who was studying the ecology of electrical charges, the then-undiscovered electron was a "fire." Freely using the word *particle,* he relentlessly experimented to learn how charges attracted and repelled each other:

> Every particle of matter electrified is repelled by every other particle equally electrified. Thus the stream of a fountain, naturally dense and continual, when electrified, will separate

> and spread in the form of a brush, every drop endeavoring to recede from every other drop.

Franklin used primitive electrical storage devices, such as long glass tubes that generated static, and Leyden jars, also made of glass, which were the batteries of the time. They delighted Franklin and enraptured his friends and members of his Junto, a Philadelphia social club dedicated to civic improvements and scientific matters. Always a practical joker, Franklin had once rigged a picture of King George II to give people a jolt if they touched his crown.

The experiments that led to the kite, the key, and the lightning rod refined mankind's knowledge of electricity and what it could do. From his playful explorations into the nature of lightning, Franklin was able to define the electric force as a "fluid thing," saving countless buildings and lives as a result.

Franklin's electrical letters were still essential reading well into the mid-nineteenth century, and his electrical experiments made him famous on both sides of the Atlantic. Walter Isaacson, in his stellar biography of Franklin, writes:

> Few scientific discoveries have been of such service to humanity. The great German philosopher Immanuel Kant called him the 'new Prometheus' for stealing the fire of heaven.... In solving one of the universe's greatest mysteries, he had conquered one of nature's most terrifying dangers.

Franklin set the standard for the adventurous solo scientist venturing into terra incognita, and Tesla doubtlessly absorbed Franklin's observations and experimental technique at an early age. In *My Inventions,* Tesla made explicit his thoughts not only on an inventor's unique and superior powers of observation, but also on the inventor's higher purpose—an obligation to improve human existence via the harnessing of natural forces:

> An inventor's endeavor is essentially lifesaving. Whether he harnesses forces, improves devices, or provides new comforts and conveniences, he is adding to the safety of our existence. He is also better qualified than the average individual to protect himself in peril, for he is observant and resourceful.... I seem to have acted under the first instinctive impulse which later dominated me—to harness the energies of nature to the service of man.

It's no wonder that Philadelphia's Franklin Institute displays a bust of Nikola Tesla (donated by the Tesla Science Foundation) front and center in its cozy electricity exhibit room.

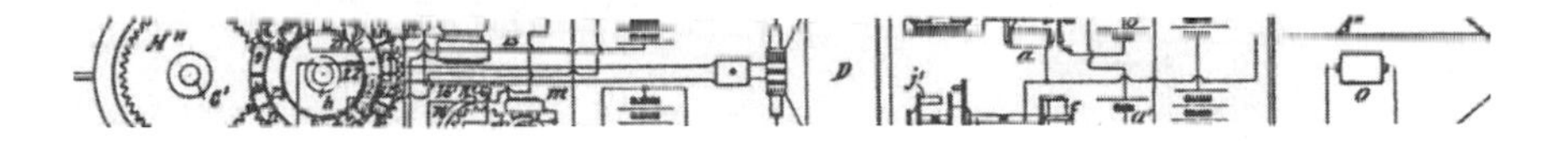

The Century of the New Prometheus

Tesla was born into a century brimming with "heretical" ideas, including the romantic notion that although God didn't directly control humanity, the creator's intentions could be understood through natural laws and science.

When Tesla was two, the great English scientist Michael Faraday effectively retired to Hampton Court after he was awarded a house by Queen Victoria (he died when Tesla was eleven). Faraday was a relentless experimenter and illuminator of the principles of electromagnetic induction—that is, how electricity is created by magnetic fields. The son of a London blacksmith whose family once had to survive a week on a single loaf of bread, Faraday started his career under the tutelage of Sir Humphrey Davy in 1813. Three years later he published his first paper. His career of meticulous experimentation would bring fresh insights in chemistry, electrochemistry, electromagnetism, and physics.

Side view of Faraday's magnetic sparking coil, 1831.

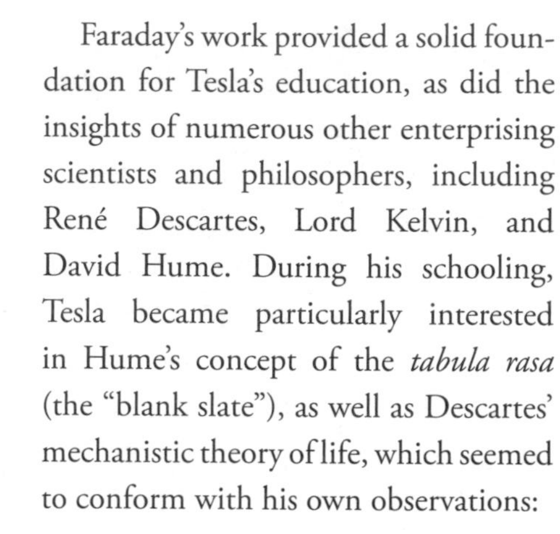

Faraday's work provided a solid foundation for Tesla's education, as did the insights of numerous other enterprising scientists and philosophers, including René Descartes, Lord Kelvin, and David Hume. During his schooling, Tesla became particularly interested in Hume's concept of the *tabula rasa* (the "blank slate"), as well as Descartes' mechanistic theory of life, which seemed to conform with his own observations:

> "The incessant mental exertion developed my powers of observation and enabled me to discover a truth of great importance.... I gained great facility in connecting cause and effect. Soon I became aware, to my surprise, that every thought I conceived was suggested by an external impression.... I was but an automaton devoid of free will in thought and action and merely responsive to the forces of the environment."

While this insight about human helplessness and free will would have turned many people cynical, Tesla—possessing extraordinary eyesight and hearing—was heartened by the idea that superior sensory abilities and an inclination toward incessant observation might give certain individuals, including himself, an advantage in the world:

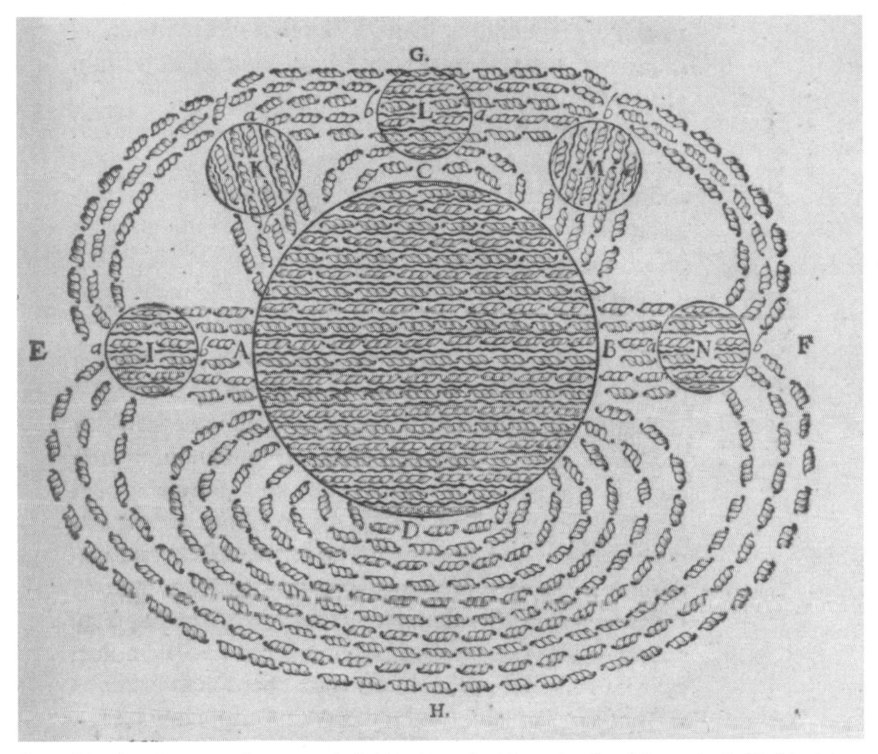

One of the first drawings of a magnetic field by French philosopher René Descartes in his *Principia Philosophiae*, 1644.

“We are automata entirely controlled by the forces of the medium being tossed about like corks on the surface of the water, but mistaking the resultant of the impulses from the outside for free will. . . . certain defects in the brain . . . deprive the automaton, more or less, of that vital quality and cause it to rush into destruction. A very sensitive and observant being, with his highly developed mechanism all intact, and acting with precision in obedience to the changing conditions of the environment, is endowed with a transcending mechanical sense, enabling him to evade perils too subtle to be directly perceived.”

Tesla's conclusion about his own "transcending mechanical sense" clearly strengthened his sense of resolve and obligation to the whole of humanity.

When Faraday's career took flight, the nineteenth century was challenging nature's dominion through the power of scientific reasoning. No longer was death considered a finite proposition in the physical sense. Could the vast energy of the sky reanimate life? As experimenters such as Luigi Galvani and Alessandro Volta observed in the 1780s, when they applied current to frog legs, it appeared that there was "animal electricity." Young Nikola had his own encounter with this electricity while stroking the back of his pet cat. In his "A Story of Youth Told by Age," written and dedicated in 1939 to the daughter of a Yugoslav ambassador who shared with him a love of cats, Tesla recalled, "Macak's back was a sheet of light and my hand produced a shower of sparks loud enough to be heard all over the place." When he asked his father what caused the sparks, Milutin considered it for a moment before responding, "Well, this is nothing but electricity, the same thing you see through the trees in a storm." This led three-year-old Tesla to wonder, "Is nature a gigantic cat? If so, who strokes its back?"

That question would animate Tesla for most of his creative life.

Volta would go on to create one of the first batteries—the Voltaic pile—a device that would occupy Faraday in hundreds of hours of experiments. The volt, a unit of electricity that would be a central part of Tesla's later vocabulary, was also named after the Italian scientist.

Another scientist whose name would become part of the everyday scientific lexicon was William Thomson, Lord Kelvin. The Irish mathematical physicist determined the lower limit of temperature (absolute zero), and in honor of his achievement the unit of absolute temperature came into standard use as the kelvin. This distinguished scientist also did important work in other areas, including electricity and telegraphy, that Tesla was eventually able to build upon:

> “In 1856 Lord Kelvin had exposed the theory of the condenser discharge, but no practical application of that important knowledge was made. I saw the possibilities and undertook the development of induction apparatus on this principle.”

Lord Kelvin demonstrates to students in the cellar he used as his laboratory in his early days at Glasgow University.

In fact, Kelvin's conductor had been formulated after he was asked to contribute his expertise to some of Faraday's experiments on the hypothesized transatlantic cable. Unprecedented progress in electricity and communication was only possible because scientists and inventors were able to build upon the work of those who came before them.

The year that Faraday published his first paper (on caustic lime)—1816—was also known as "the year without a summer." The previous year, a volcano in modern-day Indonesia called Mount Tambora had blown its top, leaving a crater nearly four miles wide. A cloud of ash spread across the world, blocking out enough sunlight to lower global temperatures. That summer, a group of English bohemians led by the poet Lord Byron sought refuge from violent thunderstorms near Lake Geneva in Switzerland. The party included John Polidori, Byron's personal doctor, as well as the poet Percy Shelley and his soon-to-be-wife, Mary Wollstonecraft Godwin. Byron had absconded from England after rumors surfaced of numerous sexual affairs, including stories of incest and general debauchery.

Since Scrabble®, Battleship®, and World of Warcraft® hadn't been invented yet, the literary friends hunkered down in Byron's Villa Diodati on the lake and challenged each other to come up with compelling ghost stories. Polidori, who was the constant subject of ridicule by Byron, penned *The Vampyre*, the inspiration for Bram Stoker's *Dracula* later that century. Basing his work on Byron's bloodsucking relationships with almost everyone around him, Polidori crafted a tale that entranced readers, although earlier anonymous versions were falsely attributed to Byron himself.

Mary Shelley, however, took the game much more seriously than anyone else. Enraptured by the endless thunderstorms set against the Alps, Shelley wrote *Frankenstein, or The New Prometheus*, which was published in 1818. The mother of all horror stories was a romantic defiance of nature itself. What if life could be rebuilt and restored through the power of lightning? What kind of creature would come of it? Shelley's "hideous progeny" was born of the electricity that Tesla sought to tame, even as a child in the latter half of that century.

Johann Wolfgang von Goethe, the great German poet who had done scientific work on colors of light, closely followed Byron's group and their work.

This relief in Dresden, Germany, shows Prometheus grabbing a lightning bolt from the sky, with a modern city as the backdrop.

Goethe, whose own tragic play *Faust* would consume sixty years of his long life, praised *The Vampyre* as Byron's "greatest ever work" while Polidori was attempting to assert his authorship of the popular book. Dogged by the predatory shadow of Byron at nearly every step in his life, Polidori committed suicide at the age of twenty-five by drinking a glass of cyanide. Luckily, Byron's daughter, Ada Lovelace, survived the monstrous ego of her father to anchor another realm of technology that would later become digital computing.

It was a bold era that presaged inventors in the romantic mold like Tesla. The undercurrent of the cultural and scientific tsunami that led to two waves of the Industrial Revolution also saw humanity playing with supernatural fire. In addition to Mary Shelley's "Prometheus," Goethe's *Faust*, which explored the moral quandary associated with the advance of technology and industrialization, also dominated the public imagination. Did one have to cut a deal with the devil to acquire power, beauty, and immortality? Tesla would wrestle with such Faustian themes for most of his life.

TeslAction 2 Be Unrelentingly Curious

According to Tesla, "Deficient observation is merely a form of ignorance and responsible for the many morbid notions and foolish ideas prevailing." The inventor clearly held the importance of observation as a central tenet in his creative pursuits, and although he had tremendous confidence in his mental prowess when it came to understanding the world around him, he also was well aware that there was much to be learned from those who came before him. But observation was at the top of Tesla's pyramid of curiosity.

All of the great minds who laid groundwork for Tesla were gifted observers. Da Vinci, Franklin, and Faraday all had a powerful grip on his imagination in childhood and early adulthood; without their attempts at understanding the relationships between natural phenomena, Tesla would have had a much steeper hill to climb. Like da Vinci, he wanted to fly; like Franklin, he hoped to capture lightning; and like Faraday, he was a meticulous experimenter. He tinkered, played, and goofed around with ideas and inventions his forebears had puzzled over for generations. As a boy, he embraced the idea of taking things apart just to see how they worked.

Tesla's relentless curiosity was a hallmark of great scientists and inventors, and the passion with which he investigated the work of others—whether experimenters, philosophers, or novelists—provided the foundation for many flashes of insight. With this in mind, think about the artists, thinkers, and entrepreneurs who inspire you the most.

- **What are your favorite works—literary, visual, or musical? Have you gathered any important lessons from them? Did any of them shape your own purpose or outlook on life? How do they speak to you?**
- **Are there any books, movies, or albums you have been wanting to read, watch, or hear? Make a list and pick your top ten, and see if you can work them into your schedule**

over the next year. Keep a notebook or tablet handy to jot down any ideas or insights.

- Which notable scientists, artists, leaders, musicians, and other luminaries—past and present—do you find most inspiring? Consider embarking on a mini research project to uncover more about their work, perspectives, and psychology. Do you notice any similarities between their outlooks and/or personalities and yours?
- Try to recall instances when you found yourself surrounded by nature or in the company of nonhuman creatures. Did anything surprise or captivate you?
- Is there any particular skill or trade you've been wanting to learn? There are lots of online courses available these days, and websites like YouTube.com contain an enormous repository of lessons and tutorials. Every college has an adult education or lifelong learning curriculum. If you just want to sample lectures, join your college's alumni association or see what's available at the local community college. They all have lecture and arts series.
- How do you enhance your curiosity? Take field trips. Go to museums. Every time I go to a big city, I try to go to a new museum. I've been exploring New York City for forty years and *still* haven't seen everything! Walk through interesting neighborhoods. Listen to music or view art that's out of your comfort zone. You can discover new things at any age. Open the window and let some fresh air in.

III

FLASHES OF LIGHT

THE PSYCHOLOGY OF TESLA

The rising sickness,
Bursts of light—
An ascension
Into the sweet secrets of angels.

—J. F. W.

The light flashes would come without warning. Bursts of images would overwhelm Tesla's psyche from boyhood through the end of his life. Possibly they were traumatic, psychosomatic aftershocks that plagued him after his brother Dane's death in 1861.

At the age of fourteen, Dane had been thrown from his father's magnificent Arabian horse. This was the same steed that had actually *saved* his father, Milutin. The horse had thrown Milutin after being spooked by wolves, but then it bolted home and led rescuers back to where the priest was lying in the snow, bruised but conscious.

A thoughtful and well-read man, Milutin would fully recover to minister to his congregation, spread out in the frontier of Smiljan, where the Austro-Hungarian Empire abutted the Ottoman Empire in Lika, a region in present-day Croatia. No one in the Tesla family, though, ever overcame

An 1880 engraving of Karlobag, Croatia, a seaside town in Lika-Senj county bounded by the Velebit mountain range. Tesla, who roamed the mountains of Croatia as a youth, was born in nearby Smiljan.

the guilt and sorrow caused by Dane's accident. The oldest son, Dane was charming, handsome, and brilliant. He was destined for great things in a part of the Balkans where borders were constantly changing and opportunities abounded in greater Europe, a continent carved up by imperial families like the Hapsburgs.

Nikola's mother, Djuka, was extraordinarily gifted, having descended, as Tesla puts it, "from a long line of inventors" in an established Serbian family. Living in a rural area torn by centuries of conflict, she *had* to be creative. She raised the food, tended the house, and ensured that Dane's and Nikola's minds were open to the new possibilities of a century exploding with ideas and revelations in nature and science. Nikola recalled,

> “My mother was an inventor of the first order and would, I believe, have achieved great things had she not been so remote from modern life and its multifold opportunities. She invented and constructed all kinds of tools and devices and wove the finest designs from thread, which was spun by her. She even planted the seeds, raised the plants and separated the fibers herself.”

After Dane died, the family couldn't bear the idea of living in Smiljan, so they moved to Gospic, a larger town with a population of 3,000. Nikola had retreated into the deep caverns of his mind, spending endless hours in his father's generous library, absorbing every subject and memorizing everything on the way to learning seven languages and mastering the principles of physics and electromagnetism, which would hold sway over him for more than eighty years.

Was there some force that could short-circuit mortality itself? The question preoccupied Tesla as he read and memorized Goethe's *Faust*, a literary masterpiece on what it means to be human. Groomed by his father to become a priest, Tesla rebelled against the clerical life as he started to learn about the emerging laws of nature. After being stricken with cholera and horrified at the thought of being unable to pursue study as an electrical engineer, Tesla impressed upon his father that he had a different path. His father relented and allowed him to be schooled at a quality secondary school in Graz, Austria, starting in 1875.

Nikola's father Milutin (**LEFT**, 1819–1879) was a stern preacher as well as an idealistic reformer, possessing a large, diverse library of books as well as noteworthy wit and memory. Tesla recalled his mother Djuka (**RIGHT**, 1822–1892) as an ingenious tinkerer and tireless worker.

By all accounts, Tesla received an excellent formal education, learning everything that the burgeoning nineteenth-century European intellectual explosion had to offer on all things scientific. The end of the Victorian era was bursting with radical notions, and Tesla was ripe to receive and transcend them. No longer was God's hand extended directly to Adam's as depicted in Michelangelo's Sistine Chapel painting. Darwin's *On the Origin of Species*, published in 1859, offered an alternative view of how we came to be human.

Darwin's ideas coursed through the Victorian age like a runaway freight train for the rest of the century (as they do even today in some dark places of the world), causing the great biologist to fear that he "killed God." According to Darwin, Nature decided which species survived, and the fittest made the cut. This thinking later permeated into the work of Darwin's cousin Francis Galton, the father of descriptive statistics and eugenics, a field of study which held that "inferior forms" of our species should be culled—dangerous ideas that were later embraced by mass murderers like Hitler and Mussolini.

From Migraines to Mind Journeys

What influenced Tesla's remarkable mental capabilities? From a young age, Tesla was plagued by "luminous phenomena" that marred his vision, seemingly at random:

> "In my boyhood I suffered from a peculiar affliction due to the appearance of images, often accompanied by strong flashes of light, which marred the sight of real objects and interfered with my thought and action. They were pictures of things and scenes which I had really seen, never of those I imagined. When a word was spoken to me the image of the object it designated would present itself vividly to my vision and sometimes I was quite unable to distinguish whether what I saw was tangible or not.... Sometimes it would even remain fixt in space tho I pushed my hand thru it."

Judging from Tesla's autobiographical timeline, these visions started in his preteen years, although he probably experienced them for the rest of his life.

Tesla denied that these were signs of mental illness, contending that "in other respects I was normal and composed" and the images were "the result of a reflex action from the brain on the retina under great excitation. They were certainly not hallucinations such as are produced by diseased and anguished minds." Although the neurological basis for these mysterious visions is not known, they bring to mind the scintillating auras that often accompany migraine headaches. Indeed, it is possible to have a " visual migraine," in which a person experiences the aura without the headache.

Interestingly, these flashes were often triggered by intense situations and in at least one instance caused acute discomfort that persisted for days:

> "They usually occurred when I found myself in a dangerous or distressing situation, or when I was greatly exhilarated... [and] attained a maximum when I was about twenty-five years old.... I felt a positive sensation that my brain had caught fire. I saw a light as though a small sun was located in it and I past the whole night applying cold

> compressions to my tortured head. Finally the flashes diminished in frequency and force but it took more than three weeks before they wholly subsided. ”

Over time, Tesla learned to recast the visions into positive sojourns, using them to his advantage. The mind, often stuck in a loop on images or experiences, can be rechanneled into more peaceful waters, and Tesla understood this.

> “ To free myself of these tormenting appearances, I tried to concentrate my mind on something else I had seen, and in this way I would often obtain temporary relief; but in order to get it I had to conjure continuously new images. ”

By his own account, the effectiveness of this method gradually diminished, forcing him to push the limits of his imagination. He visualized journeys that he would take. The cerebral sojourns would have depth and detail. He would see new places, people, and things. In an interview in 1896, he described one such "mind journey":

> “ Have you ever abandoned yourself to the rapture of contemplation of a world you yourself create? You want a palace, and there it stands built by architects finer than Michael Angelo [*sic*].... You fill it with marvelous paintings, and statuary and all kinds of objects of art. You summon fairies if you are fond of them.... Now you walk out in the streets of a wonderful city. Perchance it is one of my cities. Then you may see that all the streets and halls are lighted by my beautiful phosphorescent tubes, that all the elevated railroads are propelled by my motors, that all the traction companies' trolleys are supplied by my oscillators... ”

Clearly, Tesla's vivid imagination provided the framework for many of his future inventions, although it was only one part of the process. According to Tesla biographer W. Bernard Carlson, Tesla found that, in order to make real breakthroughs in technology,

A c. 1884 engraving of the modest village of Gracac (**ABOVE**), near the village and mountains where Tesla sought refuge in the 1870s to evade the Austrio-Hungarian Army draft, is contrasted with a twentieth-century postcard of fully electrified downtown Chicago (**BELOW**). The vivid imagination of Tesla was instrumental in making the fantasy of a highly interconnected, luminous modern city a reality.

> an insight, intuition, or hunches had to be refined in the mind through rigorous thought and analysis.... For Tesla, then, an ideal would not appear suddenly full-blown but was the result of two cognitive activities: roaming around the imagination and carefully examining possible solutions.

As Tesla became increasingly immersed in the world of science and engineering, he began to picture and rotate diagrams and machines in his mind's eye:

> “For a while I gave myself up entirely to the intense enjoyment of picturing machines and devising new forms. It was a mental state of happiness about as complete as I have ever known in life.... The pieces of apparatus I conceived were to me absolutely real and tangible in every detail, even to the minute marks and signs of wear.”

Focused visualization, which Tesla first taught himself at the age of seventeen, morphed into an easily accessed instrument for projecting new ideas. Tesla would mentally create images of machines he wanted to build and tinker with them inside his head. What did it look like? How did it work? In much the same way as Leonardo sketched his studies of nature in his notebooks or Shakespeare invented words on the page, Tesla drew what he wanted to build and attempted to perfect it—often without building a physical model. This ability is known as spatial reasoning in the lexicon of modern neuroscience.

In his 1919 autobiography, Tesla suggested that his predilection for visualization over drawing, writing, and building coincided with a preference for undisturbed thought. As such, his visualization skills were remarkably advanced, to the point of being even lifesaving. One day, for example, he was caught in a fierce current and nearly drowned:

> “Just as I was about to let go, to be dashed against the rocks below, I saw in a flash of light a familiar diagram illustrating the hydraulic principle that the pressure of a fluid in motion is proportionate to the area exposed, and automatically I turned on my left side.”

Simply by understanding the nature of moving water and visualizing its lessened impact on a smaller surface area, Tesla applied hydrodynamics to swim to safety. By his twenties, his visualization capabilities were a distinct asset as he formulated the alternating-current system that would eventually become inextricable with the astonishing technological gains of the second industrial age.

Most psychologists would call Tesla's visualization technique the core of an "intuitive" approach to creativity—that is, a method in which most ideas emerge from thinking and not from doing. It's as if Tesla had a working 3-D printer in his mind that could construct something from abstract code and images.

Tesla's eventual rival, Thomas Edison, was adept at a more "plastic" approach, by contrast, tinkering with solid objects to work through his experiments. He tried thousands of materials before arriving at the best filament for his successful incandescent lightbulb in the 1870s, while Tesla conjured images of engineering diagrams and electrical devices in his head.

Then there was Tesla's eidetic memory. Not only could he memorize entire poems, which his mother had encouraged, but he could actually commit epic poems, like *Faust*, to memory. Looking at an entire page was sometimes enough for it to stick in Tesla's memory. The psychologist Milena Bajich believes this predilection may have evolved because of where his family lived. On the edge of the Ottoman Empire, it wasn't unusual for Turkish soldiers to defile symbols of Orthodox Christianity and forbid Serbian poetry and music—a common practice in occupied countries. According to Bajich, Ottomans would come by and "piss on the Christmas roast. Dancing without music was common." So Serbian epics like *The Battle of Kosovo* were committed to memory through an oral tradition that couldn't be defiled by marauding soldiers. "Tesla,

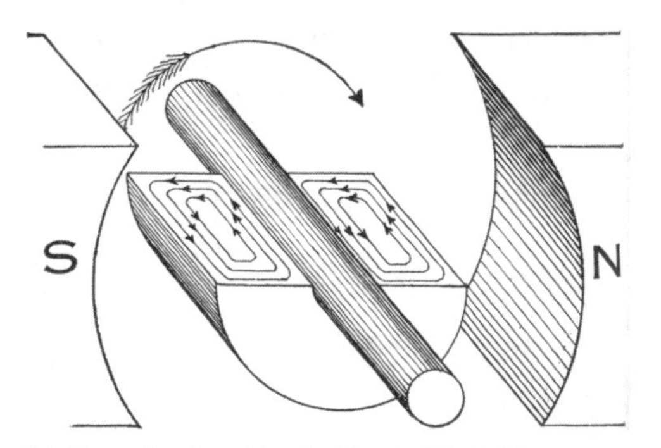

This illustration from *Hawkins Electrical Guide* (Theo. Audel & Co., 1917), which shows the eddy currents that occur when a solid metallic mass is rotated in a magnetic field, may resemble the types of diagrams Tesla rotated in his mind as he contemplated the alternating current. Eddy currents consume a considerable amount of energy and may generate a harmful rise in temperature.

due to his heritage, may have been more exposed to this type of tradition and hence used it well in supporting his other talents," Bajich notes.

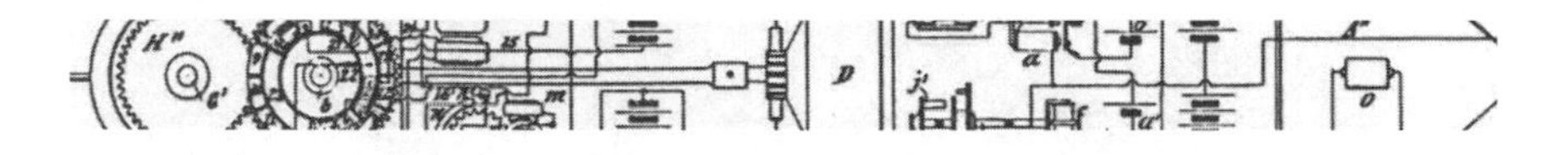

THE (EXTRAORDINARILY) CREATIVE MIND

In *Tesla: Inventor of the Electrical Age*, W. Bernard Carlson, taking note of the "wild" and "intense" sides of inventors, observes that "for the development of disruptive technologies, both activities [imagination and analysis] are needed in equal measure." Dr. Milena Bajich, who is also a painter, perhaps exhibits this dual nature herself. She has observed that creativity doesn't spring from a single skill, condition, or part of the brain: Language skills, sensory modalities, analytical skills, and life experiences all play a role—but like Carlson, she notes the tendency toward a dual nature. After analyzing Tesla and the creative attributes of others, she discerned the following:

- **Highly creative people have a great deal of energy, but they're often quiet and at rest.** Tesla spent a great deal of time in contemplation—just walking around—when he wasn't in his lab. Bajich: *"Creative people have loads of energy. They can go for hours and days in pursuit of their craft. They will throw themselves in pursuit of the 'idea' and seem boundless in this pursuit. However, they can also and do benefit from being at rest and using 'quiet' time to germinate ideas or simply allow for a 'pause' to take place."*

- **Highly creative people are both "convergent" and "divergent" thinkers.** Convergent thinking is logical and structured, while divergent thinking is more intuitive, imaginative, and free-wheeling. Although Tesla's ideas may have sprung from his divergent side, the convergent side made them into whole systems. Neuroscientists used to call this "left brain" and "right brain" thinking, but they have since come up with new theories that show how the brain works as a whole. Bajich: *"Convergent thought comes from an educated, deductive form of reasoning where a correct answer is 'deduced' from all of the learning that came before. Most people tend to hover in this area. Divergent thinkers come up with many possible solutions or a 'bunch' of plausible ideas to the same problem. Tesla clearly did both."*

- **Highly creative people are both playful and subversive, but also diligent.** As a boy, Tesla liked to experiment with his homemade popgun and pretend-fight with cornstalks, and later he enjoyed gambling and carousing, yet he also knew when to settle down and concentrate on his work. Bajich: *"The gambling was only for a short time in his youth, where he approached it like an addiction (an obsessive-compulsive style, if you will), but he was broken from the addiction by his mother giving him all of her money and saying to him, 'You will not be free until you spend it all.' . . . He was diligent with his work, for sure. He would be in his laboratory for a long time, but he also went to plays, listened to opera, and did a variety of things that were 'play' for the era."*

- **Highly creative people are rooted in reality, yet they indulge in fantasy.** Tesla's visions became working models of things that actually functioned in the world. Bajich: *"Tesla knew that certain things worked because of experimentation and because of the knowledge the society of his time already possessed, but he also was able to 'dream' of a world where things could be different and better than what he experienced."*

- **Highly creative people are often ambiverts**—that is, they are both extroverted and introverted. Tesla could be a recluse or a bon vivant. While he enjoyed the company of others and was a consummate showman, he took long, contemplative walks everywhere he went. He reveled in his own company, which often spurred his most creative moments.

- **Pride tempered by humility is often an attribute of highly creative people.** In 1894, *New York World* reporter Arthur Brisbane made this observation of Tesla: "He stoops—most men do when they have no peacock blood in them. He lives inside himself." Yet Brisbane then noted, "He has that supply of self-love and self-confidence that usually goes with success." Tesla had a lot to boast about, yet he realized there was a lot he didn't know. Bajich: *"Creatives know that they stand on enormous shoulders of those that came before them, and so it is often a man like Tesla, who with all of his abilities and accomplishments, could reflect on his accomplishments and take pride from them, but also maintain a sense of humbleness."*

- **Highly creative people often defy gender role stereotyping.** While not as extreme as the androgynous, beloved musicians David Bowie and Prince, Tesla was known for being slightly effete in his demeanor and dress. Many highly creative males and females, regardless of sexual orientation, play with the boundaries of gender imposed on them by society.

- **Highly creative people will withstand great suffering and pain in order to achieve great pleasure.** Imagine Tesla's joy at becoming the most famous engineer on the planet after getting the Niagara Falls power station running after a stint digging ditches! Tesla, like most innovators, had to run an enormous gantlet before successfully engineering the Niagara Falls plant. He had to persuade Morgan that his investment would pay off while facing the strident opposition of Edison. Few people with big ideas succeed right away. Bajich: *"The creative pursuit can lead to pain and suffering.... Such was [Tesla's] passion and pursuit, it led to much public ridicule displayed by...people who could not 'see' his vision nor understand it. People ridiculed going to the moon as a fantasy until it became a reality."*

Clearly, in order to be successful, creative individuals need to optimize their dual nature, channeling their intense energy appropriately. They need to be focused, balanced individuals who can simultaneously embrace unconventional ideas. Without a doubt, being obsessed and passionate is germane to being creative, but creators of all types must also weigh their objectives against what's possible so that they don't burn out.

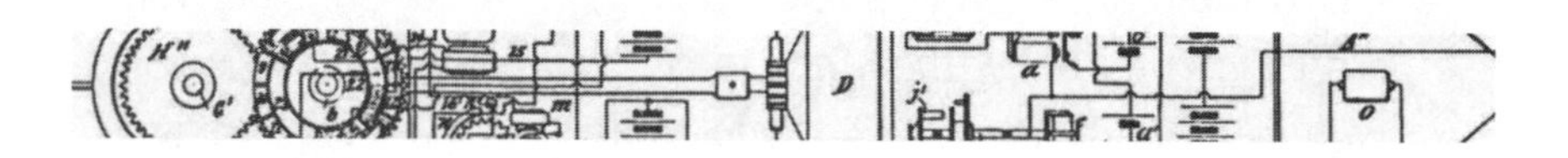

The Celestial Symmetry of the Rotating Magnetic Field

What's the fuel that drives creativity? According to psychologist Mihaly Csikszentmihalyi, who did pioneering work on the subject at the University of Chicago, you need to experience a concentrated mental state called "flow." He defines that feeling thus:

> The ego falls away. Time flies. Every action, movement, and thought follows inevitably from the previous one, like playing jazz. Your whole being is involved, and you're using your skills to the utmost.

Csikszentmihalyi, who wrote about flow in his classic book of the same name, states that you need to reach your "autotelic self"—that is, the part of your psyche that "easily translates potential threats into enjoyable challenges." That means performing, speaking, or acting; putting yourself out there for all to see; taking a chance; using your creativity as a way to reach out. To access your autotelic self, you need to set goals, become immersed in an activity, focus your attention, and "learn to enjoy your immediate experience."

Flow manifested itself for Tesla when the idea of the rotating magnetic field came to him toward the end of his formal schooling. He was having a particularly rough time, staying up late, gambling, and overworking himself in his studies, not unlike many college students today.

Tesla was struggling to improve the direct-current (DC) motor, with its perennial inefficiency and annoying commutator, which constantly sparked and needed constant replacement. A professor had told him improving or replacing the DC motor couldn't be done, although Tesla knew there was something *he* could do to solve the problem. It was one of the most pressing electrical engineering problems of the day. Everyone from New York to Prague was working on this vexing issue.

> “I started by first picturing in my mind a direct-current machine, running it and following the changing flow of the currents in the armature. Then I would imagine an alternator and investigate the processes taking place in a similar manner. Next I would visualize systems comprising motors and generators and operate them in various ways.”

Tesla's mind-visualization abilities became hypercharged when he conceived the rotating magnetic field in Budapest. A walk in Budapest's lovely Varosliget City Park at sunset with his friend Anital Szigety in 1882 triggered an unprecedented reverie that would change the world. What flooded into Tesla's mind before the picture of the magnetic field in all its glory? A section of Goethe's *Faust*, which he recited in German (here's an English translation):

> The glow retreats, done is the day of toil;
> It yonder hastes, new field of life exploring;
> Ah, that no wing can lift me from the soil,
> Upon its track to follow, follow soaring!
> A glorious dream! Though now the glories fade.
> Alas! The wings that lift the mind no aid
> Of wings the body can bequeath me.

Upon finishing the recitation, Tesla sketched in the soil of the path a rough diagram of the rotating magnetic field—the very heart of an induction

An early photograph of the scenic Varosliget Park in Budapest, where Tesla's AC epiphany occurred in 1880.

motor that needed no commutator and would power millions of factories, appliances, and the bulk of the Second Industrial Revolution.

If poetry has innate power, it's as a catalyst to allow us to see and transform other experiences more clearly. In the context of *Faust,* Goethe's troubled doctor is wrestling with his soul. He's about to do some business with the devil, who offers him power and land. It's a rough road ahead.

In Tesla's visualization, two magnetic fields dance around each other as if in a flamenco duet—jealous, passionate, both repelling and attracting—creating motion in a rotor, the shaft that rotates. At this point, you can attach just about anything to the shaft to do work: a saw blade, pulley, or wheel.

The rotating field concept begat alternating-current systems—power that could be transported great distances with acceptable loss of energy. That was in stark contrast to DC, which could only be efficiently transported a few blocks without massive loss of power. Tesla's realization that magnetic fields could forcefully interact like double hula hoops to make a shaft rotate was the engineering equivalent of all the works of Bach, Mozart, Beethoven, and Brahms—only instead of concert hall audiences having their souls collec-

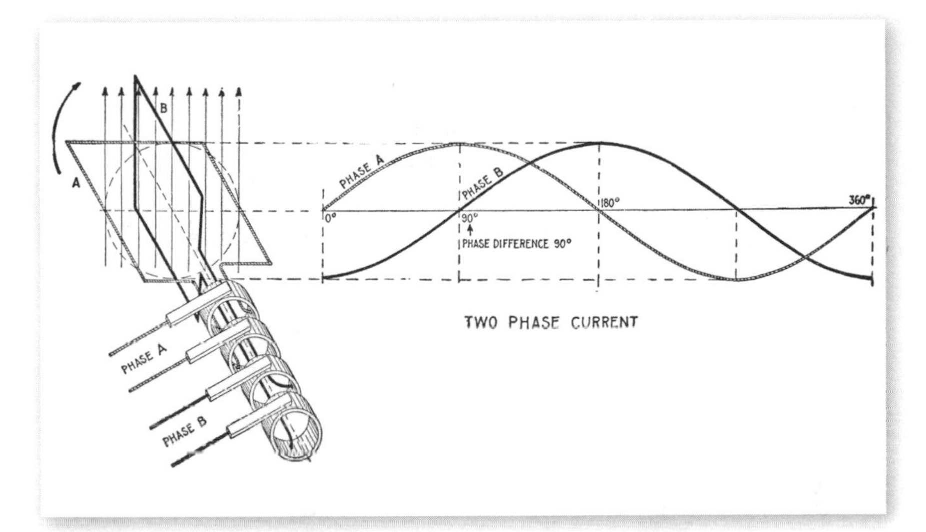

This *Hawkins Electrical Guide* figure illustrates the "dancing" two-loop alternator and sine curves of the two-phase alternating current.

tively moved, street cars could be propelled, machine tools would hum, and civilization would be able to build things faster and cheaper.

As Carlson has observed, Tesla could integrate and shift the visual with the mechanical and literal parts of his mind:

> Borrowing Goethe's imagery, the induced currents were the invisible wings that would lift the rotor and set it spinning.... Tesla chose to do the opposite: rather than changing the magnetic poles in the rotor, why not change the magnetic poles in the stator?... As we will see, this willingness to reverse standard practice—to be a maverick—was one of the hallmarks of Tesla's style as an inventor.... Perhaps he came to this hunch about using alternating currents while reflecting on Goethe's imagery of the sun retreating and then rushing forward.

How one man's poem was integrated into an idea and opened up a world of possibilities was a fulcrum of creative flow. *Metaphor became invention.* Words were transmuted into a schematic of a machine.

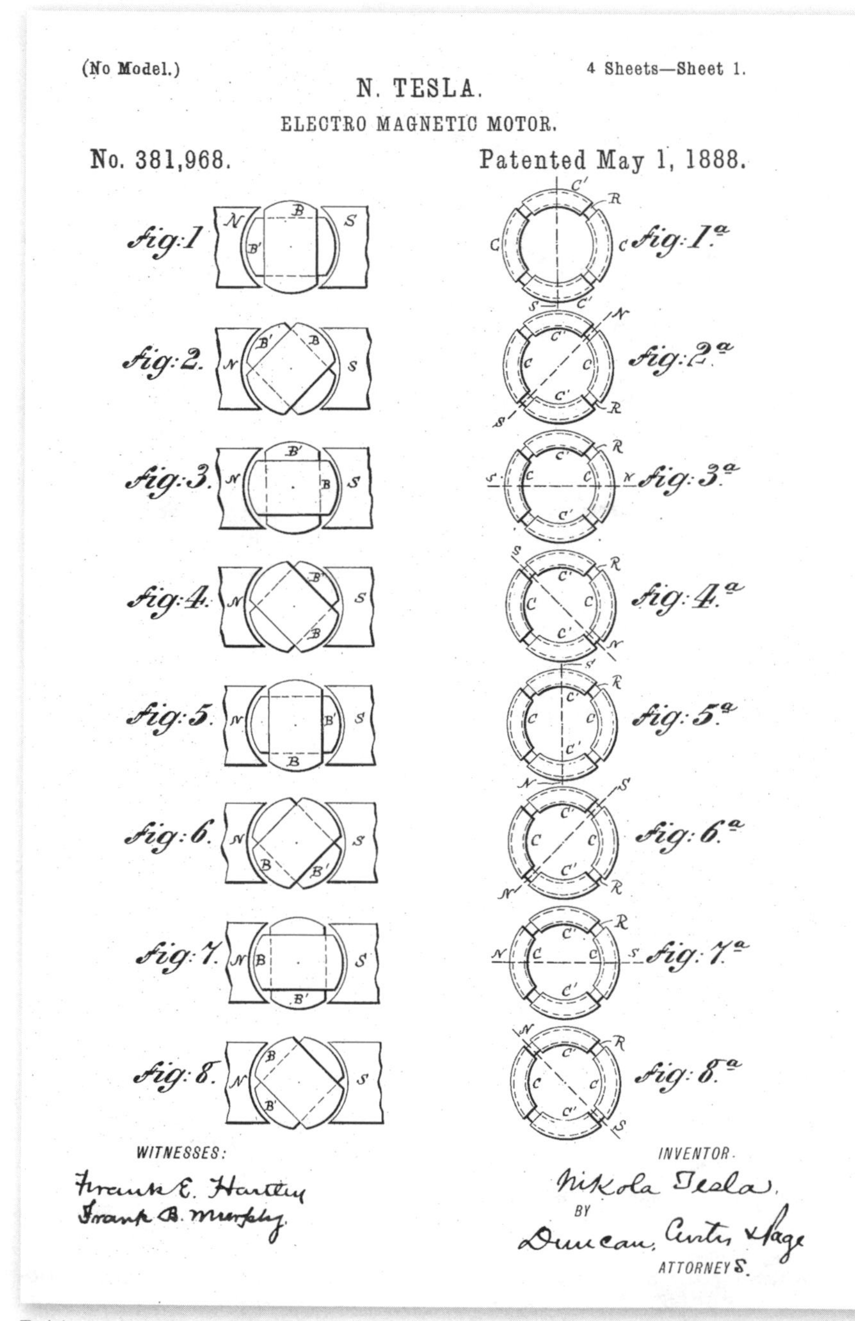

Tesla's visualization of the rotating magnetic field is illustrated in this 1888 patent diagram of his AC electromagnetic induction motor.

This illustration by German artist Hans Stubenrauch (1875–1941) shows the Spirit of the Earth appearing before the scholar protagonist who seeks infinite knowledge in Goethe's *Faust*.

Such miracles happen every day, although this day pretty much launched the bold, technological foundation of the twentieth century and is with us today in so many forms you don't notice its ubiquity. Nevertheless, Tesla's idea would take years before it triumphed as the keystone of modern electrical systems.

When I was first sorting through and analyzing the FBI file on Tesla, there was nothing revealing about his creative process, which fascinated me more than the question of whether he had created weapons of mass destruction. I wanted to know more about how he thought, how his ideas evolved, and how he overcame hurdles. He had a few obstacles to overcome, including some presented by one of his heroes: none other than Thomas Edison.

TeslAction 3 Practice Visualization

Do you need to have a profound psychophysical experience like Tesla did in order to channel ideas? Although we're always looking for epiphanies, sometimes breakthroughs come one little stone at a time. As Carlson notes, there are two sides to an inventor: the "wild" side where imagination reigns and the "intense" side dominated by rigorous analysis. Tesla leveraged both sides in his discovery of the rotating magnetic field.

Visions come in all kinds of forms—some of them painful—and they can be controlled. As I noted earlier, Tesla was able to channel these involuntary flashes of light into "mind journeys" that took him to other places and cultivated his imagination. Over time, he began to focus his mind more squarely on engineering concepts and developed the ability to exercise an exceptional degree of control over these mental manifestations. But the real breakthroughs happened when he was in a relaxed "flow" state, revolving objects and schematics in his mind with remarkable clarity.

How do you get into a flow state? For some, the answer is music or art. For others, it could be bowling, bridge, or knitting. It's the activity that engages both body and mind—and often the spirit. Going to a religious service does it for billions of people engaged in a faith or spiritual practice. And then there's yoga, tai chi, the martial arts, swimming, running, and taking strolls in nature. The infinite variety of human experience provides countless ways to experience flow. The key is to do something active and be thoroughly engaged in it in the moment. (Watching cable television or texting probably won't do it.)

Once, when I had suffered a neck cramp so intense that it felt like someone was touching me with a cattle prod for four hours, I engaged in guided imagery to relieve the distressing tension. I envisioned body surfing in the Pacific off the coast of Costa Rica, which I had done on the last days of a blissful family vacation several years earlier. I heard the spider monkeys as I emerged from the rain forest, then ran along the coarse volcanic sand into the waves. In body surfing, you use your frame as a surfboard, allowing the swell of the ocean to lift you up and glide you onto the beach. I felt the warm water caressing me, along with the ultimate soft landing as the wave expended its energy.

Because memorable experiences are associated with images, sounds, smells, and other sensations, you can use them not only to override unpleasant emotions or sensations, but also to visualize creative solutions. Recall from chapter I that Tesla probably had an unusual condition known as synesthesia, the ability to mix sensory experiences, which is more common among highly creative individuals. Whether or not you're a synesthete, conscious cross-linking of sensory modalities can be very useful when it comes to inventing things, including prose or poetry. Before I write, I need to "hear" the words I'm going to write in my mind's ear, so I use my auditory center as a lathe to reshape my experiences into pieces of prose (ideally).

Think about Tesla turning his troubling visions into little journeys when you are about to do something challenging. You can go there, too. In TeslAction #1, you were prompted to reflect on your life experiences, personality, and goals. Now take those memories and insights about yourself to create something new. As an exercise, paint a vivid picture in your mind of a moment in your life when you felt completely at peace. What were you doing, and what were your surroundings? Call to mind the smells and sounds.

Now, visualize something simple you want to do—a recipe, art project, or sports activity, for example—and ask yourself the following:

- **What does it look like when you are doing it?**
- **How will you feel (good, I hope)?**
- **Are there any smells, sounds, or textures connected with it?**
- **If you're creating something, what does it look like? Can you draw it in your head? Make it move? Turn it around?**

If you concentrate hard, you'll be surprised at the kind of virtual reality your brain provides without any silicon-based computing involved.

IV

FROM PRAGUE TO PITTSBURGH

THE ROVING ELECTRICIAN

While Edison did much to advance technology, no invention of his has an important place in modern society. His incandescent lamp is rapidly being replaced. There is nothing yet imagined that will replace Tesla's AC motor, a key element of the modern world.

—Robert Curl, Nobel Prize Winner in Chemistry

With the image of the rotating magnetic field dancing vividly in his mind, Tesla fully devoted himself to the mysteries of physics and engineering. He wanted to share this vision, but while Edison was gaining fame in America's cultural and financial hub, relishing his triumph of engineering the incandescent lightbulb and selling it to the world, Tesla was in Eastern Europe, gambling by night and arguing by day about how his AC motor could work.

In earlier studies at Graz Polytechnic in Austria, while reading Voltaire and playing with a Gramme dynamo—both a generator and a motor—Tesla was perturbed by the highly inefficient device, which sparked and sputtered. Convinced he had a better design that eliminated the need for the sparking commutator brushes, Tesla had argued with his teacher, Professor Jacob

This 1912 illustration from *Popular Science* imagines future electric cities in which power for all needed functions in society is controlled remotely by a single lever.

Poeschl, who effectively humiliated him by delivering a lecture on why Tesla's idea would never come to fruition.

"Mr. Tesla will accomplish great things," Poeschl began, "but he certainly will never do this. It would be the equivalent to converting a steadily pulling force, like that of gravity, into a rotary effort. It is a perpetual motion scheme, an impossible idea."

Yet the seductive beauty of the rotating magnetic field made intuitive sense to Tesla; it was a shimmering image, like the Lady of the Lake in Arthurian legend. It beckoned him. As he noted in his autobiography,

> "Instinct is something that transcends knowledge. We have undoubtedly certain finer fibers that enable us to perceive truths when logical deduction, or any other willful effort of the brain, is futile."

Although he initially deferred to Poeschl, there was something of a rebel in Tesla. Just as he had defied his father in refusing to become a priest, taking up gambling, and spending hours reading Goethe and Voltaire, Tesla defied his teacher by clinging to his vision of the AC motor. He grappled with its mechanical details: How did it look when it was running? How would the magnetic lines of force weave in and out like some dynamic stitching? How would the motors work with generators, delivering the electrons like threads of energy? At first he thought these engineering problems were "unsolvable."

The Gramme dynamo, designed by Belgian inventor Zénobe Gramme in 1871, was the first machine to be used industrially to generate power. This photograph shows a hand-cranked model from the 1870s intended for use in laboratories.

In 1878, Tesla dropped out of school and left Graz, without telling his parents. His friends thought he had drowned in the Mur River,

This 1880 postcard shows Maribor on the banks of the Drava River in Slovenia. Tesla dwelled here after he dropped out of Graz Polytechnic in 1878.

but in fact he had taken up residence in nearby Maribor (in present-day Slovenia), where he got work as a draftsman and gambled nightly in a pub called the Happy Peasant. Within a year, however, he was arrested for being a "vagrant" and deported back to his hometown of Gospic.

In April 1879, Tesla's father died from illness, and in January 1880 Tesla ventured to Prague with money from his maternal uncles, but he had arrived too late to attend university classes. At the time, the baroque splendor of the Bohemian capital was one of the crown jewels of the Austro-Hungarian Empire, a city adorned by domes and the glistening Danube.

"The atmosphere of that old and interesting city was favorable to invention," Tesla remembered. "Hungry artists were plentiful and intelligent company could be found everywhere." But Tesla, feeling guilty that his education was costing his remaining family dearly, didn't stay in Prague for long. In 1881, he went after a job opening in Budapest on one of Europe's first telephone exchanges, but the telephone job didn't materialize right away, so he took a position as a draftsman in the city's Central Telegraph Office, probably at a meager salary. Nevertheless, it was a smart move, setting the stage for his jobs at Edison Machine Works in New York City. Despite the mundane work and low pay, Tesla was grateful for the mentoring he received in Budapest:

> "When natural inclination develops into a passionate desire, one advances towards his goal in seven-league boots.... The knowledge and practical experience I gained in the course of this work was most valuable and the employment gave me ample opportunities for the exercise of my inventive faculties."

Having impressed his superiors in Budapest with his math and systems design skills, he received a referral to work in Paris in the fall of 1882. Although it was another dead-end job with a less-than-cutting-edge company, it provided an opportunity that Tesla would seize to eventually cross the Atlantic.

The Continental Edison company that Tesla joined in 1882 was, in fact, Thomas Edison's European subsidiary, then managed by the sharp Englishman Charles Batchelor. A combination executive, engineer, troubleshooter, and talent scout, Batchelor was Edison's man in Europe, and he liked what he saw in the young Serb who worked diligently to improve Edison's telephones with better amplifiers for the low-volume devices. In those days, Edison's machines were plagued with technical problems, often breaking down and catching fire, so employees like Batchelor were more like repairmen than high-level engineers. When Batchelor realized the range of Tesla's skills, he sent him all over France to fix Edison's flawed machines.

Charles Batchelor, one of Tesla's early advocates.

Tesla, a bon vivant in the City of Light, loved living in Paris. Every morning, weather permitting, he swam the Seine twenty-

This 1883 photograph by Hippolyte-Auguste Collard (1811–1887) shows the Sully Bridge extending over the River Seine, where Tesla went for daily swims as a young man.

seven times in a circuit and walked an hour to his office in Ivry-sur-Seine. At night, he played billiards with his associates. As Tesla reflected, "I led a rather strenuous life in what would now be termed 'Rooseveltian fashion.'"

In 1883 Tesla traveled to Strasbourg to fix a railroad station lighting plant that had short-circuited, motivated by a promised bonus from the company if he could get it running again. "There were bacteria of greatness in that old town," Tesla observed, charming Strasbourg's mayor as he got the plant running. Yet the trip offered Tesla another opportunity: to pitch his AC motor to the town's burghers, who put up some money for a prototype, which apparently ran fairly well in early 1883. The Alsace, however, wasn't ready to embrace the technology that would light up an entire century, so Tesla returned to Paris, where he was told by Edison managers that his bonus would not be forthcoming and he should stop hawking his AC dream.

Batchelor, who admired and felt sorry for Tesla, had other plans for the inventor. He recommended that he go to New York to work directly for Edison, who was sorely in need of an ace repairman. For Tesla, it was like an invitation to ascend Mount Olympus and work for Zeus. He hoped to sell Edison on AC systems. Edison could not only give Tesla his blessing, but also provide Tesla with something that wasn't readily available in Europe—that is, access to the infinite wealth of J.P. Morgan's cartel through the powerful Edison brand value.

The Pearl Street Jewel

New York's unfolding fist of creative might in the early 1880s was mostly concentrated in lower Manhattan, the thumb of the universe for capital and innovation. Wall Street was then the home of J.P. Morgan and his web of oligarchs, henchmen, and financiers. Only blocks away was Edison's office on lower Fifth Avenue, as was his skunk works, then known as Edison Machine Works, which had moved from Menlo Park, New Jersey, after the lightbulb business took off.

Having been pickpocketed of his steamer tickets and most of his money and luggage, Tesla somehow made it across the Atlantic and through immigration. He had only four cents in his pocket when he arrived in New York City on June 6, 1884. And what he saw—the year before France gave the Statue of Liberty to America—was disorder and ugliness. There were no stately domes or palaces like those he had seen in Prague and Budapest. New York's skyline then was low-rise, jagged, and smoking from a thousand coal fires. Wires hung over the streets of lower Manhattan like dense, gnarly jungle vines, and they would crackle and often explode at random. The power lines brought high voltage to the city's arc lamps—illumination so harsh, bright, and inefficient, you could go blind just gazing at them.

Arc lamps were the best effort at the time to mass-illuminate large spaces, a technology invented by Sir Humphrey Davy, Faraday's mentor, early in the nineteenth century. Inside the tenements and office buildings, gas lamps dimly flickered, because arc lighting was too bright for indoor use. Gas was an even greater menace, though. When the flames went out, the coal gas could asphyxiate people or cause catastrophic explosions and fires. Everyone, including Edison, despised gas.

Edison had yearned to put the gas companies out of business by replacing gas with safe, incandescent lighting, but there were some Herculean obstacles to overcome: How would Edison not only deliver power and efficiently wire entire cities, but transport DC power over long distances? Tesla had the blueprints in his brain to answer that question when he walked into Edison's office and introduced himself, his letter from Batchelor in hand.

Before he met Edison, Tesla would probably have first met with Samuel Insull, Edison's young English secretary and manager of his chaotic business affairs. Like Tesla, Insull came to New York upon Batchelor's recommendation, after having run Edison's London office. Insull was even poorer than Tesla; his father was an impoverished London dairyman and itinerant temperance preacher. Yet there was something that drew Insull to Tesla at first glance. Both men had superior skills in organization and communication.

The Edison Machine Works, located on Goerck Street in lower Manhattan, 1881.

This c. 1886 lithograph shows the dense and sometimes dangerous tangle of telephone, telegraph, and power wires on Broadway in New York City.

BROS
DIAMOND
MOUNTINGS
AND FINE
JEWELRY
L. SAUTER
& CO.
RINGS
BLISS & MASON
L. SAUTER & CO.
J. B. BOWDEN
J. B. BOWDEN & CO
194
HAMILTON
&
HAMILTON JR
176
1
J. B. BOWDEN
& CO.
MANUFACTURING
JEWELERS
RINGS
MARX
&
WEIS
180
MARX & WEIS
DIAMONDS
DENNISON
FOSTER & BAILEY
ROLLED PLATE CHAINS
MAIDEN LANE
WM. BARTHMAN
WATCHES DIAMONDS AND
FINE JEWELRY
POTTER
&
BUFFINTON
RICHMOND
& CO.
176
CROUCH
&
FITZGERALD
TRUNKS
SAMPLE CASES
HAMILTON & HAMILTON JR.
FINE ROLLED PLATE CHAINS
J. B. RICHARDSON
& Co.
ROYAL
BUTTON

Nikola Tesla, New York City, c. 1885.

Although the compact, beady-eyed Insull was a largely uneducated teetotaler who didn't gamble, he was an efficient multitasker and organizer who introduced some order into Edison's helter-skelter business, which included the Machine Works, electric light companies, and other interests. Insull also was known to outwork the hyperworkaholic Edison. Insull, who would maintain contact with Tesla for the rest of his life, later acknowledged Tesla's "very great contributions of the first rank" in his memoir. Had Insull possessed the money and understood Tesla's AC technology, he might gladly have given it to Tesla to develop his system that day.

Tesla came at exactly the right time for Edison, who was struggling to keep the lights on in the lower Manhattan financial district. At the time,

Thomas Edison listens to a wax cylinder phonograph at his laboratory in Orange, New Jersey, 1888.

Edison was buying some valuable public relations by lighting up not only J.P. Morgan's office, opposite the New York Stock Exchange, but the newsroom of the *New York Times*, which sang Edison's praises some two years earlier when Edison's Pearl Street Station came online.

Standing nearly six and a half feet tall, with chiseled cheeks and dark hair, and wearing an elegant topcoat and bowler, Tesla looked the part of an aristocrat when he stood before Edison, having walked to his office from the pier. Edison, a rumpled, graying, half-deaf Midwestern scrapper with almost no formal education, looked over Batchelor's letter and couldn't help but be impressed, although he didn't want Tesla to get a big head. The brusque Edison wasn't interested in a new electrical system and told Tesla so. He needed a glorified repairman—somebody to fix the constantly short-circuiting dynamos. Well-heeled clients were getting antsy because his DC system wasn't running as advertised. If he couldn't keep the lights on, the money would dry up, which was a perennial problem since Edison's companies were so poorly capitalized.

One of Tesla's first jobs for Edison was to power the lighting on the S.S. *Oregon*, one of the world's fastest steamships at the time and the first to have its own electric lighting system. Tesla was immediately sent off to New York Harbor to diagnose the problem and get the lights working consistently. After several sleepless nights, he accomplished his task and headed back to Edison's office at five in the morning, as Edison and Batchelor (long since summoned back to New York from Paris) were just heading home. Edison glared at Tesla walking down the street, thinking he was out partying instead of working.

"Here's our Parisian, running around at night," Edison said to Batchelor derisively. After Tesla calmly explained how he had fixed the generators on the *Oregon*, he politely tipped his hat like a gentleman and walked on. Edison, suddenly recognizing Tesla's fierce dedication, then admitted to Batchelor that the young Serb was "a damn good man."

The insane hours that Tesla would work on behalf of Edison that year were mostly spent at the Pearl Street Station, which was one of the first central power stations in a big city. Although the dynamos were constantly breaking down, Pearl Street was the epicenter of Edison's planned empire. Central stations, spread throughout cities, would power neighborhoods beginning in the 1880s. Edison could charge not only for lightbulbs, which cost a princely sum of one dollar apiece—a fortune at the time—but also for the electrons that lit them up, thanks to his invention of the electric meter. All city residents could have power—*if* they were willing to pay for it.

So Edison sent men like Insull, who became something of a marketing maven, to cities across the country to sell Edison DC systems. When cities agreed to buy Edison equipment, they bought into the whole package—dynamos, wiring, streetlights, and later streetcars, which were pulled by horses, who had an occasional tendency to drop dead in the middle of the street and a chronic tendency to drop tons of stinking manure, at least before central station power took hold in cities across the world.

The most glaring problem of Pearl Street and other Edison plants is that they were inefficient on an epic scale. The dynamos were driven by pulleys connected to steam engines, which were fed by coal. The noise and smoke were horrendous. Even worse, small stations like Pearl Street could run wires only within a few blocks.

DC power couldn't go far without losing most of its initial energy. Edison had to dig channels in the streets to bury the wires and seal them with a tar-like material that didn't weather well, and electric current would often shock unsuspecting carriage horses when the wire insulation degraded. And accessible electrons then were a rich man's luxury: J.P. Morgan, who had a steam engine/generator in the basement of his Murray Hill mansion, was constantly pestered by neighbors who complained of the racket and pollution. Morgan would assuage them by sending over boxes of premium cigars, then conveniently vacate to Europe to buy more art.

Edison Pearl Street Station workers lay the tubes for electric wires in the streets of New York City, 1882.

Electricity in one's home during the Gilded Age was like owning a Raphael—only a handful of elites such as the Morgans and Vanderbilts could afford their own mini-power stations in the days before the Consolidated and Commonwealth Edison utilities provided universal power to cities.

The opening of the Pearl Street Station on September 4, 1882, was akin to the introduction of the first digital computer. Although the 100,000 feet of wiring buried in the streets were powered by an unreliable, twenty-seven-ton "jumbo" dynamo, they revolutionized the way electricity could be delivered and consumed. From one remote source, Edison provided power for safe lighting twenty-four hours a day, plus everything that was to follow in terms of machine tools, heating, elevators, and appliances.

Tesla did what he could with Edison's archaic steam-age technology, designing "24 types of standard machines...which replaced the old ones" along the way. "For nearly a year my regular hours were from 10:30 A.M. until 5 o'clock the next morning without a day's exception." As an incentive, Edison flippantly said he would pay Tesla $50,000 if he could overhaul Pearl Street so it would operate consistently without burning up. After working nonstop for nearly a year, Tesla delivered and came to Edison to claim his bonus. Out of spite, or perhaps thoroughly intimidated by Tesla's élan and engineering prowess—qualities that he himself did not possess—Edison barked in his flat Midwestern voice that he would not pay up, much less give Tesla a real raise or promotion, retorting, "When you become a full-fledged American you will appreciate an American joke." Edison reportedly offered Tesla a modest hourly rate increase, but his humiliation of Tesla dug far too deep. Insulted and enraged, Tesla resigned on the spot. He'd take his ideas where they were appreciated, although he didn't know where that might be at the time.

Out of the Ditch

Before Tesla quit his engineering job with the Wizard of Menlo Park, Edison had given him plans for arc-lighting technology, which didn't interest Edison, who was committed to his DC/incandescent business plan. In December 1884, New Jersey investors Benjamin Vail and Robert Lane sought out Tesla to ramp up an arc-lighting company. Tesla willingly climbed out of the ditches to join the start-up.

Although arc lighting was clearly an outdated and inefficient technology, Tesla, who relished the idea of improving it using his own designs, built an outdoor lighting system in Rahway, New Jersey, in 1885. Still harboring the dream of developing his AC technology for widespread use, the inventor formed his own company, Tesla Electric Light & Manufacturing. His modest capital wasn't enough to sustain the company, though, and it was soon overshadowed by the Thomson Electric Company, a business nemesis of Thomas Edison that was also backed by Morgan. Realizing that Vail and Lane had no interest in anything other than the soon-to-be-obsolete arc lighting, Tesla moved on, worthless stock certificates in hand.

Broke, bedraggled, and living in a rathole, Tesla took a job digging ditches in the streets of lower Manhattan for two dollars a day, working with a gang of other European immigrants, in 1886–87. The idea of AC power was still percolating in his mind and had circulated to others in the closely knit community of electrical engineers and financiers. His foreman heard Tesla talking about his system and quickly realized that the engineer didn't belong in the ditches, breaking his back like a plow horse. But then, Tesla had a way of escaping that backbreaking labor. He had invented a thermoelectric motor and was only too happy to talk it up with his foreman, who introduced Tesla to the investors Alfred Brown and Charles Peck.

Brown was an executive at the Western Union telegraph company. Peck, a lawyer from Englewood, New Jersey, had invested in a telegraph line that linked Washington, D.C., and Chicago in 1879. They created a rival to Western Union called Mutual Union, which quickly became involved in lawsuits and stock manipulation driven by the rapacious Wall Street swindler Jay Gould. Peck and Brown knew how to do start-ups outside of the Morgan circle, and they wanted Tesla.

Paying Tesla a salary of $250 a month—a princely amount at the time—Peck and Brown organized the Tesla Electric Company in April 1887. In addition to working on his AC motors, Tesla agreed to work on improving other technologies. They set him up in his own lab, hoping he would make them rich with his pyromagnetic motor, which was driven by heat. Although

Tesla was ultimately unable to perfect that technology, the investors believed in him, so he set out to develop his AC system.

With Peck and Brown, Tesla was not only designing, tinkering, and experimenting every day; if he conceived of a great, pragmatic idea, he filed for patents. Although the idea of AC transformers was embraced in Europe by mostly Italian and German engineers (the Siemens Company was akin to the General Electric of Europe)—a development closely watched by American engineers—Tesla wanted to be the first to synthesize a complete system that utilized his AC motors and delivered multiphase current wherever it was needed.

Joining the growing fraternity of electrical associations was the first step in Tesla's coming out as a major inventor. The American Institute of Electrical Engineers (AIEE) was by far the most prestigious, followed by the National Electrical Light Association and the Electrical Club New York. These were the bastions of aspiring men who made up the Silicon Valley of the late nineteenth century, all competing with each other. There were Edison's men, Thomson-Houston's men, and various observers who reported back to Morgan. George Westinghouse—an engineer, executive, and inventor himself—was also a member of this fraternity, although being Pittsburgh-based, he was a bit of an outsider.

By May 1888, Tesla was invited to present his AC system to the AIEE. The title of the lecture was unassuming: "A New System of Alternate Current Motors and Transformers." The audience, which included the brilliant inventor Elihu Thomson, who had an AC motor in development, was conquered by the depth and advanced nature of Tesla's system. Although Thomson challenged Tesla at the end of his talk, it was clear that Tesla had the superior product—no nasty, sparking commutators needed. For some reason, though, Thomson's companies passed on Tesla's patents, which left an opening for Westinghouse.

As with Edison and his lightbulb, there were competing inventors when it came to the AC motor. For example, a physics professor named Galileo Ferraris had built a small prototype of an AC motor in 1885 but didn't publish his results until 1888, and he was unsure if his design was practical. Reading the Ferraris paper, Westinghouse was curious as to whether Tesla had a better design, so he sent his top electrician Oliver Shallenberger to see a demonstration of Tesla's motor in his Liberty Street lab. After a brief visit, Shallenberger was convinced Tesla had the technology everyone else desired—and it worked.

Some tense negotiations ensued between the Westinghouse camp and Peck and Brown. In the end, they secured one of the most potentially lucrative deals any inventor had ever received: a royalty of $2.50 per horsepower for every AC motor, $50,000 in notes, $25,000 in cash, and guarantees that royalties would rise from $5,000 in the first year to $15,000 each succeeding year. The Westinghouse AC deal was the equivalent of buying all the basic plans to the personal computer, the mouse, and displays. Such an arrangement had the potential to make Tesla richer than Morgan, Carnegie, and Vanderbilt combined—*if* his system was universally adopted and Westinghouse stayed solvent. But there was a slight catch, which Tesla didn't object to at first: He had to join Westinghouse's engineering team in Pittsburgh to get his designs into production. Tesla had to be part of a team.

One of the first models of Tesla's AC induction motor, which he demonstrated to the AIEE in 1888.

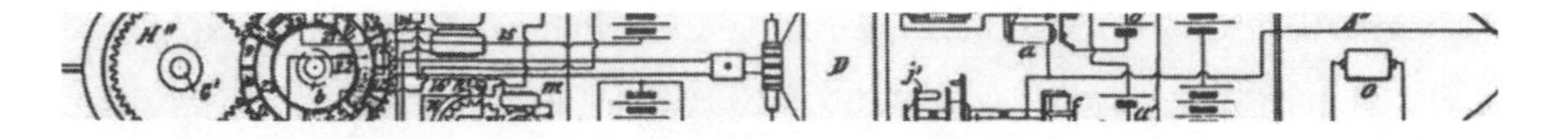

THE "BATTLE OF CURRENTS": TESLA V. EDISON

Tesla was the field marshal in what became known as the "Battle of Currents." Edison, through the Edison General Electric Company, was desperately trying to malign Westinghouse's AC systems. The Wizard of Menlo Park enlisted Insull and publicity man Harold Brown in a massive propaganda campaign to show the world how dangerous and evil AC was in comparison to the relatively "safe" DC current. Edison staged electrocutions of animals such as dogs and elephants with AC current—gruesome displays designed to show the perils of AC. When, on August 6, 1890, William Kemmler became the first person to be executed in the electric chair, Edison

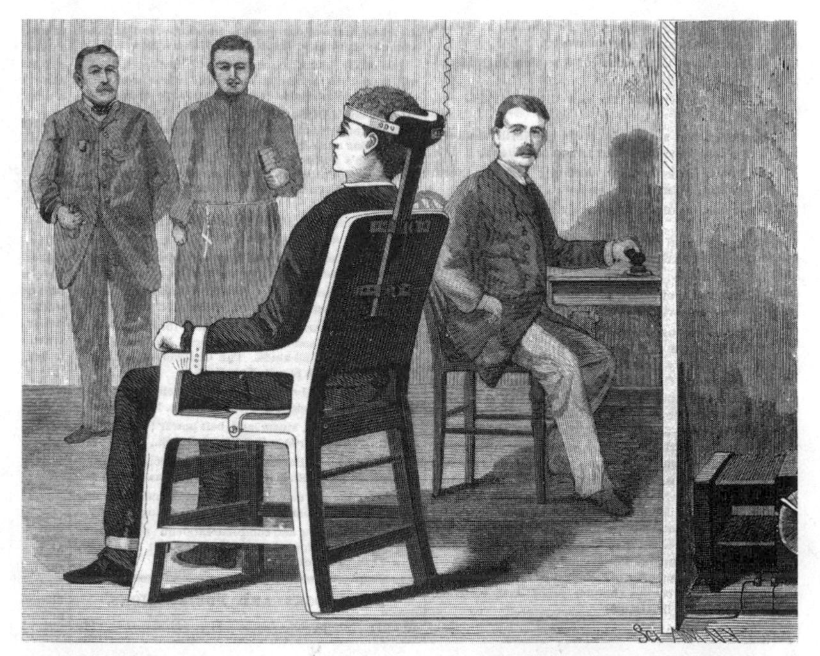

This illustration from *Scientific American* (June 30, 1888) depicts the newly approved form of capital punishment, the "electric chair," which famously utilized AC current on August 6, 1890, to execute William Kemmler.

had been so successful with his anti-AC campaign that his description of AC's electrocution horrors became a verb: "westinghoused."

The grisly execution required several jolts of current, which created a blue flame behind Kemmler's neck, set his clothes on fire, and created a burnt-flesh stench. Kemmler did not die immediately; he had effectively been "roasted," according to Jill Jonnes in *Empires of Light*. The witnesses, one of whom had fainted, were appalled. Contacted for comment after the electrocution, Westinghouse quipped "they could have done it better with an axe."

During his tenure with Westinghouse, Tesla was a whirlwind of invention while Westinghouse engineers perfected his AC system for mass distribution. Although he had been summoned to Pittsburgh to launch the polyphase project, he had little interest in staying there, shrouded in a haze from its many steel mills.

Living the luxurious life of a consultant, Tesla still yearned for his small New York lab, where he was working on high-frequency current experiments. Along the way, he invented his eponymous coil, a device that magnified current thousands of times to produce fibers of electrons that could reach across a room. While the Tesla coil had no practical use in the Westinghouse scheme, it became the basis for other experiments.

Tesla discarded more ideas than he pursued in his lower Manhattan lab. He apparently stumbled upon what later became known as X-rays but moved on to what he regarded as more important matters than "radiant energy." It was in those days that Tesla believed that electricity could be transmitted without wires at a certain frequency, and he demonstrated the effect in St. Louis. Tesla's schematic for a "wireless transmitter," which he patented but didn't develop, was seen by Guglielmo Marconi, a young Italian inventor.

Westinghouse, laboring mightily to raise capital and sell Tesla's AC system to cities across the continent, was facing long odds on survival. Outside of the Morgan sphere, Westinghouse had slim prospects of attracting the kind of money he needed to go global,

This photograph from an April 1895 edition of *Century Magazine* shows electrical sparking between condenser plates of a large Tesla coil during an experiment in the inventor's laboratory workshop.

even though he had the superior system. His biggest competitor was Thomson-Houston, which had also adopted AC but was profitable and moving steadily into manufacturing.

The weaker player in the industry, at least according to Morgan, was Edison General Electric, which was now being shepherded by financier Henry Villard, a German-born railroad magnate and friend of Morgan who wanted to expand and capitalize Edison's operations with German money from Deutsche Bank. Morgan, always focused on control and consolidation that would reduce or eliminate competition of his holdings, had a much simpler plan: Combine Thomson-Houston with Edison General Electric. Westinghouse now was competing with a goliath.

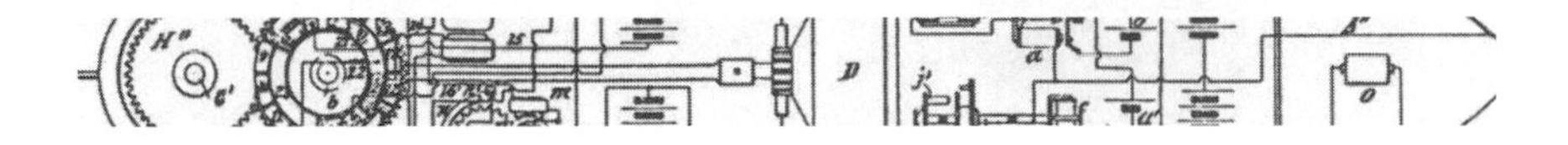

One of the first alternating-current applications was high in the Rocky Mountains in Telluride, Colorado, far away from the nexus of capital and innovation in New York City. Mining financier L.L. Nunn wanted electricity for his gold-mining operation, so he financed a power plant that ran on hydropower. He funded the dam-and-dynamo operation and sent the AC electricity to his mine about three miles away in the spring of 1891. The above photograph shows the interior of the power house at Nunn's hydroelectric plant near Ophir, Colorado. Although Tesla's system would later capture the global spotlight in Chicago and Niagara Falls, this small-scale project attracted the attention of those watching the bitter Westinghouse-Edison "Battle of Currents."

Tesla: The Outlier

As Tesla admitted years later, even though he was finally free to develop his vision of AC power, he had some significant challenges to overcome to make his technology compatible with the needs of Westinghouse:

> "My system was based on the use of low frequency currents and the Westinghouse experts had adopted 133 cycles with the object of securing advantages in the transformation.... [M]y efforts had to be concentrated upon adapting the motor to these conditions. Another necessity was to produce a motor capable of running efficiently at this frequency on two wires, which was not an easy accomplishment."

Nevertheless, the inventor persisted, and his collaboration with the Westinghouse team yielded a power system that was more powerful and pragmatic than the one his former employer had devised.

In the end, Edison was essentially left in the dust by Tesla's partnership with Westinghouse, which probably enraged Edison. Although he had sent Insull

An 1888 catalog advertisement for Pittsburgh-based Westinghouse Electric Co. and its revolutionary AC system.

to Schenectady, New York, to start a colossus of a manufacturing empire that would later become General Electric, the stream of Edison's groundbreaking innovations would soon slow down as Morgan got a tighter grip on his company and realized that DC power was becoming obsolete. Moreover, Edison himself had bet big on the wrong horse with his DC systems, even though Insull had begged him to put AC equipment into production. Tesla, the man that Edison had once humiliated, was on his way to inventing the twentieth century, although not without a few unforeseen missteps.

It would have been easy to dismiss Tesla in the 1880s and early 1890s because he was going up against Edison, Morgan, and the electricity "establishment" with some radical ideas on how to produce and transport power. In every cultural sphere, new ideas are rarely welcomed with open arms. As David Kent, author of *Tesla: The Wizard of Electricity*, points out,

> Many of his ideas seemed to be impractical or even delusional or irrational.... Then again, perhaps all great inventors become great inventors because of their willingness to let their visionary ideas fly beyond rationality. Edison, Tesla and virtually all others made many jaunts down the wrong road before finding the path to the future.

The grand irony is that Tesla would win the Edison Medal decades later in recognition of his early work on AC power. The solid gold medal, like many of his documents, was later reported missing, as I discovered when I scoured through the FBI file.

TeslAction 4 Be Indispensable

In addition to being extraordinarily gifted and curious about the physical world, Tesla became well known for his tireless work ethic. As he himself admitted, "I am credited with being one of the hardest workers and perhaps I am, if thought is the equivalent of labor, for I have devoted to it almost all of my waking hours." Apparently this was yet another quality he shared with his pragmatic, highly inventive mother, who worked from dawn until dusk in the Croatian countryside. Indeed, studying the career trajectories of some of the most successful people in recent history, author Malcolm Gladwell noted that those who "made it" invested at least 10,000 hours in their craft. The Beatles, for example, played all-night shifts in the clubs of Hamburg before they recorded their most groundbreaking songs. Similarly, Tesla had literally to get in the trenches and do exhausting, menial electrical work before he made a name for himself. While working for Edison, he put in eighteen-hour days.

To succeed at anything, you need to embrace the necessity of "learning the ropes." It doesn't matter what the work is—making music or designing rockets: All complex knowledge systems require that you develop skill sets and a great deal of practical knowledge. Hands-on experience is essential to garnering support and investment from others who can help you move up the ladder or gain fame. Tesla was very prudent in placing himself in the right environments for success. After pleading with his father to send him to engineering school, he wandered from Graz to culturally vibrant Prague and then to cosmopolitan Budapest, where he found fulfilling work and set himself up for jobs in Paris and New York under the acclaimed Thomas Edison. In addition to amassing important practical knowledge in these metropolises, he found company in other intellectuals, artists, and engineers that nurtured his unconventional ideas about electricity.

Starting at the bottom is never an easy task—it can be bruising to the ego, taxing on the body, and extremely time-consuming. You may have to kiss up to some people you don't particularly like. In some industries, you might have to work for no or little money when you first start out.

When I was a cub reporter, I was sent to the poorest communities to cover city council meetings. You build knowledge through experience and being around people who have mastered skills. Seeking mentors is a key part of this process. I usually had older editors and reporters around to tell me what to ask for—and what to avoid. That's why internships and apprenticeships are still around. Get out in the world and make yourself indispensable to those who share your goals or possess the expertise you wish to have. Put in the time to learn what works and what doesn't, whether it has to do with technical problems or client relationships. Here are a few questions to guide you as you devise your creative strategy:

- What kinds of skills will put you at an advantage in your field or make your product or creation superior to others on the market?
- Who are the best individuals or companies in your field, and what techniques, training, or creative philosophies do they possess that you might like to acquire? What kinds of questions would you ask them if given the chance?
- Are there any "hubs"—particular cities or neighborhoods—where people or companies with goals or values similar to yours are located?
- Think about where you want to be one, five, and ten years from now. What things would you have to accomplish in order to meet your goals in each of those time frames? Are any academic degrees or certifications required? Would it make sense to relocate to a new city?
- What are the unsavory aspects of your endeavor? Are you prepared to undertake them? If not, do you have a plan for delegating them? What risks are you willing to take?
- What kind of funding or investment do you need to meet your goals? Do you have a plan for acquiring it?

V

THE WIZARD OF ELECTRICITY

TESLA'S LIGHTNING SHOW

A dance with oblong bulging lines—
A hora weaving in and out of the center,
Pulsating, alive with symmetry,
Dervishes disciplined by nature.

These fleet forces drive a shaft,
Compelled by invisible tendrils,
Hava nagila in a metal cylinder—
Irrepressible joy contained and ever revolving.

—J. F. W.

STANDING NEARLY SEVEN FEET TALL in thick, cork-soled shoes, Tesla walked onto the Chicago World's Fair stage and stared at the audience with the deep mystery of a master illusionist. Then, when a switch was pulled offstage, Tesla's gaunt body became a lightning rod attracting some 250,000 volts. Sparks flew from the tips of his fingers. Thanks to his insulating shoes, Tesla wasn't electrocuted, but the effect was beyond dramatic. It was an otherworldly—even gaudy—demonstration of the power of electricity, and when combined with Tesla's spinning Eggs of Columbus, it made for a world-class spectacle that delighted fairgoers.

Tesla demonstrates a trick in which he holds balls of flame without burning his hands in this illustration from the *Pearson Magazine* article "The New Wizard of the West" (May 1899).

One of Tesla's Eggs of Columbus displayed at the Westinghouse exhibit in Chicago, 1893.

To convince Peck and Brown that his AC ideas were worthy of their investment, Tesla had employed a parlor trick that would become world famous: He created a device that made a copper egg stand on its end. It was based on a probably apocryphal story that Christopher Columbus made an egg stand on its end to convince Queen Isabella to underwrite his voyage to the New World. In Columbus's ruse, he slightly cracked one end of the egg. When Tesla reprised the trick, his metal egg began to rotate on a disk without any visible influence. The rotating magnetic field was the invisible hand. Tesla got his funding (as did Columbus) and would revisit the trick to demonstrate his technology to millions of people only a few years later. I was lucky enough to marvel at the simple charm of Tesla's original Eggs of Columbus apparatus when it came in a traveling exhibit to Chicago in 2011 from its permanent home at the Tesla Museum in Belgrade.

At the fair, Tesla was fully transformed into a chimeric icon—a mythic creature exuding flame. Standing erect as a statue in his elegant black tailcoat with fire bursting out of his fingertips, the handsome scientist projected an Olympic quality. More importantly, Tesla had taken his presentation skills to

Tesla radiates electricity in this 1894 illustration from the *New York World*.

a new level. Small lab demonstrations now became theater for hundreds of thousands of people who had never seen the raw power of alternating current. Tesla not only invented the system; he literally embodied its power. He was one with the force he was harnessing.

While Tesla was generally a loner in his design process and preferred to work with just a handful of assistants, Peck and Brown had convinced him that he needed to focus on ways to promote his ideas. As investors and entrepreneurs, they needed some influential people to see what the inventor had to offer. In this effort, they were not disappointed: Not only could Tesla synthesize his ideas through visualization, math formulas, and building prototypes; he was a nascent showman who could beat the drum to draw wide audiences. By 1893, the reserved, erudite inventor was embracing his inner P.T. Barnum to sell his power systems with great success.

Of all of the displays at the 1893 World's Columbian Exposition in Chicago, Tesla's was the one that was guaranteed to inflame the imagination of the more than 27 million attendees. Yet Tesla wasn't there as a circus sideshow. He was the prime-time salesman for his polyphase system and the Westinghouse manufacturing empire that stood behind it. All that was efficient and useful about AC was part of the Westinghouse exhibit (albeit on a tiny scale), which was directly competing with Edison's General Electric's display, an obelisk-like tower of smallish lamps, for the grand prize: the electric operating system of the twentieth century.

Next to the rival General Electric display, Westinghouse Electric & Manufacturing Co. advertises the "Tesla Polyphase System" at the World Columbian Exposition.

Ultimately, the fair was the endgame in the AC/DC war. Westinghouse was financially imperiled due to shrinking capital and a national economic malaise. The Pittsburgh titan needed a moonshot to sell his system. Congressmen, senators, and captains of industry would be attending.

More importantly, mayors of large cities, desperate to power trolley systems, light their streets, and vanquish harsh arc lighting and dangerous gas lamps, were anxious to see which technology would prevail in keeping 250,000 bulbs lit. Chicago, most of all, would be a prime beneficiary, not just from hosting the fair and tens of millions of attendees, but from whatever technological innovations emerged to make it and hundreds of other cities safer, more habitable, and more civilized.

Fair planners Daniel Burnham, the great urban planner and architect, and Frederick Law Olmsted, the legendary landscape architect and designer of New York's Central Park, dubbed the fair the "White City," but in reality most of Chicago was enveloped in filth, disease, and poverty. Edison's and

Westinghouse's lights promised to bring Chicago—and every other urban area—out of the dimly lit and dirty reality of the nineteenth century. The fastest-growing city in the world was celebrating the 400th anniversary of the "discovery" of North America by Columbus (a feat so controversial, dubious, and genocidal that few would think of dedicating a World's Fair to it today).

Like a phoenix, Chicago, having risen from the ashes of a devastating fire some twenty years earlier, was pushing hard to reinvent *every* city in 1893. Burnham and Olmsted had a heroic mission: to transform a swampy downtown bog of some seven miles into a complex of palaces heralding a brand-new age of spectacle, luxury, and efficiency to the entire world. Featuring an enormous Ferris wheel (the first) and hundreds of sideshows, the fair was to be festooned with canals, electric gondolas and trams, and neoclassical buildings—the largest exhibition of the soon-to-be-modern world.

Occupying one square mile with more than 200 buildings, the fair was a city of dreams, what Burnham and Olmsted imagined as part of their "City Beautiful" movement that stressed extensive planning, generous green space, and buildings that looked like transplants from Rome or Greece—on massive doses of steroids.

Chicago, then and now, was a city of strident contrasts. With the idealized city emerging on its South Side, Chicago's environment was a toxic miasma. The water supply, Lake Michigan, was perennially polluted by human waste and the effluent of the slaughterhouses, since the open sewer known as the Chicago River flowed directly into the lake. Factories relentlessly fouled the air and water. Steam engines pulling passenger trains rolled right into downtown, spewing coal ash and cinders. Every street was covered with soot and horse manure. The rapidly growing population was nearly 80 percent foreign-born, and the hundreds of thousands of immigrant workers, profiled in Upton Sinclair's *The Jungle*, battled cholera, typhus, tuberculosis, and short life expectancies in the dangerous mills and plants girding the city. One visitor, the British writer Rudyard Kipling, who couldn't wait to get out of town, said Chicago was "inhabited by savages."

Yet there was something about Chicago, with its burly charm, that made it an ingénue, though one with a dirty face and torn clothes. The ragamuffin had some spunk and aspired to great things.

This World Columbian Exposition advertisement shows a "grand birds-eye view" of the spectacular grounds and buildings, designed by Daniel Burnham and Frederick Law Olmstead, that comprised the so-called "White City" in 1893.

Acres.
44 Acres
Woman's Building.
United States Gov't Building.
Illinois State Building.
Fisheries Building, 2 Acres.
Gallery of Fine Arts, 5 Acres.
U.S. Naval Exhibit.
COLUMBIAN EXPOSITION AT CHICAGO, ILLINOIS, 1892-3.
OF THE DISCOVERY OF AMERICA BY CHRISTOPHER COLUMBUS.
New York.

This photomechanical print of Haymarket Square jammed with commercial wagons shows a filthier side of Chicago in the 1890s.

The University of Chicago, which was only about three years old when the fair came to town, abutted the White City's "Midway Plaisance." The college, which would eventually be home to numerous Nobel Prize winners, wanted to be everything the city wasn't: educated, informed, progressive, and world-class.

The new Chicago was the stand-in for L. Frank Baum's Oz. Young men with ideas, including Frank Lloyd Wright, Clarence Darrow, Carl Sandburg, and Theodore Dreiser, learned their trades in the Emerald City. Women like Jane Addams and Ida B. Wells fought for social justice; Susan B. Anthony championed women's rights. Antonin Dvorak heard the rhapsody of the New World, while Scott Joplin would introduce a new kind of music called ragtime, opening the way for America's indigenous music, jazz, to flow into Chicago like a summer storm. Swami Vivekananda, who influenced Tesla's pan-spiritualism, brought the novel message to the fair's "Parliament of World's Religions" that the world could live in harmony.

Newfangled inventions like gasoline-powered cars, box cameras, safety razors, and typewriters made their debut at the Chicago fair. Consumer products that would become household names and remain so today—Juicy Fruit® gum, Aunt Jemima® syrup, Cream of Wheat® and shredded wheat, Pabst beer, Cracker Jack®—heralded the beginning of a national consumer culture and the advertising industry that created it.

In the war of the currents, the staging was less prosaic. The World's Fair was the Stalingrad for Edison and Westinghouse. Like Edison, Westinghouse had bet the house on the fair and its massive national stage to sell AC to a public that had little idea that you could even pipe power into factories or homes.

GE had bid high on the contract to illuminate the fair, which was the largest venue to be lit up by electric lighting at the time. Westinghouse bid low to get the contract, but it came at such a cost that it put his company on the brink of bankruptcy at a time when the country was in a recession and all the companies in the electrical industry were fighting patent wars in the courts against each other.

Crowds sport umbrellas outside the fair's Electricity Building.

The Museum of Science and Industry in Chicago, the last remaining original building from the World's Columbian Exposition, is now the largest science museum in the Western Hemisphere. Although the buildings were largely built to be temporary—and most burned down after the fair—the one remaining structure still heralds the progress of civilization. In the midst of researching this book, I visited the museum, a temple of science and progress with its triple jumbo Palladian domes, obscenely large halls, Corinthian columns, and quirky exhibits on everything from hatching chicks to a captured Nazi submarine. I strolled through the museum's backyard, placid with its lagoon in Jackson Park and wooded island, which had hosted a Japanese pavilion during the fair. For a place that once welcomed tens of millions of people with everything from fan dancers to electric thrills, the museum was remarkably serene in the middle of the turbulent South Side.

As the "Battle of the Currents" drew to a close, an entirely new world had been born in Chicago, spurred by the efforts of some 65,000 exhibitors. Some 700,000 daily visitors who came into Chicago's vital rail hub saw all of this, plus exhibits from exotic countries like Egypt and the revolutions emerging in agriculture, commerce, and the arts. The theme, then and now, was doggedly American: *Everything* could be done better. People could live better lives through agriculture, automation, education, mass production, science, and electricity.

Westinghouse Goes Over the Falls

A year before the Niagara contract went to Westinghouse, Sam Insull, Edison's confidant and chief operating officer of sorts, reluctantly advised his boss to consolidate his electrical companies with J.P. Morgan's interests. Morgan then assembled the General Electric Company in 1892 and took Edison's name off the door. Insull didn't care much for Morgan, but he saw the consolidation as a route for the graceful survival of Edison's legacy; Insull realized that Edison was not going to win the "Battle of the Currents" with DC. It also didn't polish Edison's brand that the Cataract Construction Company, Morgan's entity behind Niagara Falls, had rejected Edison's plan to wire the facility with DC equipment and bring the power to New York City.

Many who knew Edison saw Insull's move as a betrayal, although Edison and Insull remained friends until Edison's death in 1931. In any case, Edison continued as director of the new General Electric Company, was awarded millions in stock, and was free to tinker in his manufacturing labs, one of which was located in sunny Fort Myers, Florida. Yet Edison's halcyon days as a bootstrap inventor became part of his permanent legend, burnished by friends like Henry Ford, who lionized Edison in a museum near Detroit (Greenfield Village), creating the modern iconography of all things Edison.

Competing against the giant GE conglomeration, Tesla's polyphase system was Westinghouse's only slingshot with one projectile remaining, and it was a massive one: wiring the power plant at Niagara Falls. Although a 108-mile AC line in Germany was one of the world's first major electrical systems, it commanded relatively little attention in 1890. In contrast, Niagara, given its proximity to Buffalo, New York, presented a much higher-profile stage.

Flush with the success of the fair, Westinghouse beat out General Electric for the contract to build the power station at Niagara Falls. At the helm of the project, which once had on its board Edison himself, was Tesla, the chief engineer.

In one of his many visions as a boy, Tesla had dreamed of tapping the power of the falls, where the waters of Lake Erie plunge over the Niagara escarpment into the Niagara River and eventually Lake Ontario.

ABOVE: Mr. and Mrs. Westinghouse stand in front of the mighty Niagara Falls. **BELOW**: By the turn of the century, Niagara's might had been harnessed by Westinghouse and Tesla's massive dynamos to power nearby Buffalo, New York.

"I told my Uncle that I would go to America and carry out this scheme. Thirty years later, I saw my ideas carried out at Niagara and marveled at the unfathomable mystery of the mind."

On November 16, 1896, when the power was switched on, the kinetic energy of water turned Tesla's three 5,000-horsepower AC dynamos, powering Buffalo's electric street trolleys twenty six miles away. Though that moment sent hundreds (and later thousands) of horses out to pasture, it was the vindication that Tesla and Westinghouse sought beyond the showmanship of the Chicago fair.

To historian Henry Adams, the generators were a symbol of infinity, providing ever-greater amounts of captured lightning that allowed civilization to flourish. H.G. Wells, who would base much of his greatest work on the grandiose mystery of such power, wrote in 1906 of Tesla's dynamos:

> They are will made visible, thought translated into easy and commanding things. They are clean, noiseless, and starkly powerful. All the clatter and tumult of the early age of machinery is past and gone here; there is no smoke, no coal grit, no dirt at all.

From the time Edison opened his Pearl Street Station in New York to the switching-on of Niagara's massive dynamos, the entire industrialized world was on its way to being electrified by central station power plants, thanks to Edison, Tesla, Thomson, and Westinghouse.

By 1893, there were more than 2.5 million electric lights in use powered by 3,500 isolated power stations. Although it's not as clean as Wells would have you believe—coal-fired and nuclear plants are undeniably polluting—Tesla's system is still largely the dominant technology in most electrical grids in the early twenty-first century.

SCIENTIFIC AMERICAN

[Entered at the Post Office of New York, N. Y., as Second Class matter. Copyright, 1896, by Munn & Co.]

A WEEKLY JOURNAL OF PRACTICAL INFORMATION, ART, SCIENCE, MECHANICS, CHEMISTRY, AND MANUFACTURES.

Vol. LXXIV.—No. 14.] ESTABLISHED 1845. NEW YORK, APRIL 4, 1896 [$3.00 A YEAR. WEEKLY.

NIAGARA FALLS POWER PLANT—ONE OF THE 5,000 HORSE POWER TURBINE WHEELS.—[See page 215.]

An 1896 edition of *Scientific American* features one of Tesla's 5,000-horsepower dynamos.

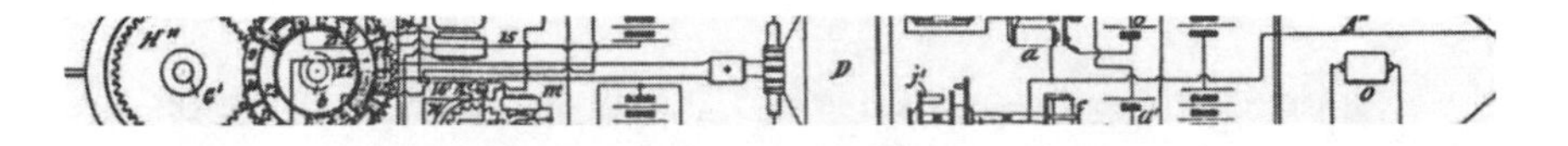

TESLA'S SOCIAL CIRCLE

Back in New York, Tesla backed away from power systems to explore more exotic things. He socialized, stayed at the posh Waldorf Astoria, and invited luminaries of the day to his lab for the boiled-down demonstrations he had offered in Chicago.

Although Tesla's later image suggested a crazed recluse, the Tesla of the closing days of the nineteenth century was a cultured, charming European polymath. Invited into the cultural, creative circle of Robert Underwood Johnson, publisher of *The Century Magazine*, and his beautiful wife, Katherine, who adored Tesla, the scientist would melt his small audiences with poetry readings, discussions of philosophy, and displays of what electricity could do.

Johnson recalled a woman in his social group (whom he doesn't identify) asking Tesla what he did for a living. Suave and modest, Tesla responded, "Oh, I dabble a little in electricity."

LEFT: The Waldorf Astoria, c. 1900. **RIGHT**: The dashing bachelor-inventor-showman Nikola Tesla at age thirty-four, 1890.

"Indeed!" the woman exclaimed. "Keep at it and don't be discouraged. You may end by doing something some day."

Tesla didn't disappoint his guests when he wanted to engage them in the wonders of the electric. For the great environmentalist John Muir, he could illuminate a lightbulb in his hand without the use of wires. The famed architect Stanford White, designer of the old Madison Square Garden, probably was intrigued by the structure of Tesla's devices, towers of power as they were. Ever the impresario, Tesla would demonstrate for his friends a Zeus-like mastery over the forces of nature—a power that could either help bring about world peace or create weaponry that would wreak unparalleled destruction.

Tesla had come across the work of the great American writer Mark Twain during his bout with cholera in 1873. Even though the doctors had given up on young Nikola at that point, Twain's literary wit somehow managed to revive him:

> **"One day I was handed a few volumes of new literature unlike anything I had ever read before and so captivating as to make me utterly forget my hopeless state. They were the earlier works of Mark Twain."**

When Tesla met Twain in person, "I told him of the experience and was amazed to see that great man of laughter burst into tears." The two men became good friends, bonded by their shared love of storytelling.

In one memorable instance, when Twain insisted on "feeling" Tesla's nonlethal, low-frequency current, Tesla turned on the juice. At first, Twain felt a soothing calm, followed swiftly by an urgent surge of incontinence that made him bolt to the water closet, much to the amusement of the other guests. It was Twain, perhaps more than anyone else in the Johnson circle, who succumbed to the chimeric inventor's quest. Although he was the best-known American writer

Tesla looks on as Samuel Clemens (a.k.a. Mark Twain) holds a loop over the resonating coil, passing high-tension current through his body that brings the lamps to incandescence. Photograph reproduced in *The Century Magazine,* April 1895.

on the planet, the creator of Tom Sawyer and Huckleberry Finn was in constant financial peril. Hoping to enjoy a steady income stream, he invested in various inventions and companies that had a propensity to go bankrupt. One invention was a typesetting machine that didn't quite work well enough and nearly bankrupted Twain, who had to book a worldwide speaking tour to pay off his debts. Twain also came up with and patented a history game called The Memory Builder. After going bust on an investment, Twain either hit the road to travel or made some easy money telling his stories on the lecture circuit, which was quite lucrative in those days.

For many people of that era, the age of invention was a road to newfound wealth, although it never materialized for the man who convinced Ulysses Grant to write his autobiography (perhaps one of the finest ever written) while dying of cancer, "just so that he could leave something behind for his family." When Twain was abroad, his fame ensured that he would be received like royalty—no small feat for a Civil War deserter and newspaperman from

Hannibal, Missouri. Presenting himself as some kind of financier, in Vienna toward the end of 1898, Twain offered to promote "that destructive terror which you [Tesla] have been inventing":

> Some interested men might be discussing means to persuade the nations to pair with the Czar & disarm. I advised them to seek something more than disarmament by perishable paper contract—invite the great inventors to contrive something against which fleets and armies would be helpless & thus make war thenceforth impossible.

Twain was likely referring to Tesla's remote-controlled boat, and it's clear that the prescient Twain was interpreting the simmering tensions between the great powers that would lead up to World War I. The two men would remain friends for some twenty years, until the author's death in 1910.

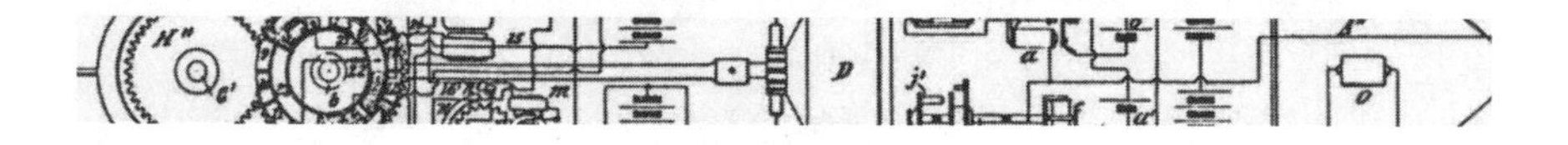

The Creative Breakthroughs of the Post-Niagara Years

The World Columbia Exposition may have celebrated Tesla's new system and sold it to the world, but the massive event was only a midpoint of Tesla's long creative life. He probably could have been noted in history as the greatest electrical engineer who ever lived, but he wasn't concerned with posterity—at least not as a technical titan.

As his chimeric transformation continued, he grew metaphorical wings to take his ideas to another level. What enraptured Tesla in his cozy lab only a few blocks from Edison's early workshops was a desire to take the power he could create and broadcast it around the world. His oscillators and magnifying transmitters could not only create enormous amounts of energy, but also send it through the atmosphere in discrete channels, which is how the electromagnetic spectrum is broken up for today's microwave and radio transmissions.

Whatever secret inventions Tesla created in his last Manhattan lab didn't survive. In 1895, a fire of unknown origin destroyed most of his research and equipment. While investors were quick to provide funding for future research, Tesla had moved on to a philosophy of power: What could it do? How should it be applied?

By 1898, having mastered wireless transmission of electricity on a small scale, Tesla demonstrated the remote use of power in a demonstration at Madison Square Garden. He sent a tub-shaped vessel with tiny antennae motoring around an artificial pond, and it heralded the birth of robotics and remote applications. Behind the toy boat was something even more powerful:

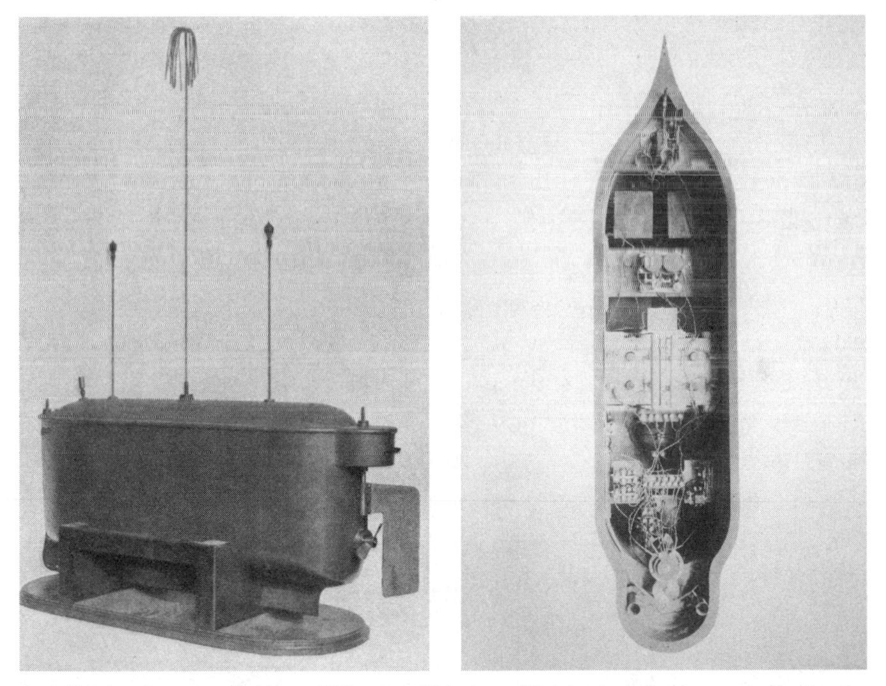

LEFT: Tesla's radio-controlled boat, 1898. **RIGHT:** This view of Tesla's telautomaton reveals the circuitry involved in remote-controlled robotic technology, which brought science one step closer to the practical modern computer.

a radio transmitter that could work on many channels. Everything from TV remote controls to deadly military drones emerged from this entertaining if simple demonstration.

Still, Tesla was restless as he opened another lab in lower Manhattan. Something more Promethean kept him up at night, and it wasn't the invention and perfection of radio or remote control. He considered electrical energy the source of human thought itself, and he pondered the existence of mental telepathy and "transmission of mind efforts" in a letter to Westinghouse.

An autographed photo of Swami Vivekananda, the Hindu mystic who inspired Tesla and many others, Chicago, 1893.

After his meeting with Swami Vivekananda in the 1890s (although there are conflicting accounts on when or if the meeting happened), Tesla's thoughts leaped to the idea that human energy was an energy field in the universe that resonated throughout the earth and sky. Resonance—the way in which nature vibrated, much like a tuning fork or a violin or piano string—was the key. Match that resonant frequency, and you send energy vast distances, although this concept would have little commercial value at the time while threatening entrenched interests. To the moneyed elite bankrolling General Electric, Westinghouse, Marconi, and other growing industrial combines, Tesla must have become a bête noire when he wrote in 1896:

> "The end has come to telegraph and telephone monopolies with a crash. Incidentally, all the other monopolies that depend on power of any kind will come to a sudden stop. The earth currents of electricity are to be harnessed. Nature supplies them free of charge."

While Tesla could have created his own manufacturing empire, he seemed content experimenting in his small labs through the end of the nineteenth century. Unfortunately, he frequently undersold his enterprise. He sold his Tesla Electric and Manufacturing Company to Westinghouse Electric for a measly $5,000 and 200 shares of Westinghouse in 1888. The down payment for the purchase was $950.

Observers have argued over the years that Tesla was perhaps more devoted to works for humanity than to commerce, which is a subtext in the weapons research he conducted that later caught the attention of the FBI and defense establishment. In all his letters to Westinghouse (that is, to the founder of the company and his successors), Tesla had a noble purpose in mind: He wanted to build a universal system that would benefit—and later *protect*—everyone. Those sentiments would mean little to G-men and other shadowy government entities who saw enemies everywhere. Meanwhile, the men who had become multimillionaires on Morgan's consolidations through his many tightly controlled trusts from steel to electrical equipment were hearing a discordant melody in Tesla's words.

While the great minds of the late nineteenth century waltzed into their twilight years, Tesla's transformation into a "threatening" chimera was just beginning. No longer would he be the charming inventor at cocktail parties or the delightful eccentric dining alone in Manhattan's finest hotels. As the dawn of the twentieth century approached, he was off to broadcast energy across the planet and through the earth.

TeslAction 5 Sell Your Ideas

While it's romantic to think of the lone inventor struggling in a small, remote laboratory, the leaders of innovation have many collaborators. Without the willingness or ability to effectively sell one's ideas, however, such collaboration isn't really possible on a large scale. A lot of creative people get tripped up by the need for salesmanship, as it requires a totally separate set of skills that can leach time and energy from innovative pursuits. However, it often makes the difference between success and failure.

The creation of the modern digital computer involved thousands of people, starting in World War II with geniuses like Alan Turing, who built upon the groundwork of nineteenth-century thinkers like Ada Lovelace and Charles Babbage. Even the vaunted Apple team behind Steve Jobs had to be able to translate Jobs's vision into workable, everyday appliances. When Tesla had his greatest success, he had backers and engineers transforming his AC system into components of a larger system. Without Peck, Brown, Westinghouse—and even his ditch-digging foreman—Tesla's inventions might never have become commercial realities.

In seeking collaborators, know what kind of talent you need. Some people's skills lie in the financial area while others have technical or marketing talent. While all collaborators should believe in your overall mission, the most effective enterprises have collaborators with a diverse array of opinions on how such a mission can be achieved.

Once you have the image in mind of what you want to create, you have to synthesize it, making it as real as possible to potential collaborators and/or investors. If it's something technical, you need blueprints and diagrams. For written pieces, an outline and draft will do. For visual projects, think of a "movie" trailer that will get the concept across in a few seconds. Depending on how detailed your vision is, you'll need to flesh it out in order to communicate it effectively to others. When I'm doing a lecture, for example, I develop an extensive series of notes and major points. I abhor PowerPoint, so instead I rely upon old-fashioned white/black boards and flip charts with permanent markers. Everything is planned and rehearsed: During a lecture, I don't read anything directly and mostly ignore my notes after I have my outline in mind.

While it's great to have some Eggs of Columbus to entrance your audience with a little magic, cogent visual and verbal explanations of how your project will work will win over your audience. Stories definitely help: Psychological studies have demonstrated that they make information easier to memorize. By extension, they can cement in people's minds how your ideas fit together and why they are important. And as Carlson notes in his Tesla biography, "No ideal, no idea, no invention goes anywhere unless one is willing to tell a story about it, a story that another person finds interesting and persuasive."

Go beyond the standard PowerPoint slides. Tailor your pitch to your audience, and make eye contact. Spin your project as the best solution to whatever your audience is in need of. Make a personal connection with your listeners. Jokes also help to break down boundaries between the speaker and the audience.

So the next time you have to meet with an important client or potential collaborator or make a presentation, ask yourself the following:

- **How can I make my pitch both informative and entertaining? Can I get my audience laughing, perhaps?**
- **Why should my audience care about my project? What can it do for them?**
- **Can I entrance my audience with a demonstration or visual? Can I create something of a spectacle?**
- **Why is your idea important? How does your vision fit into the context of bigger and more ambitious plans? Drive that point home.**

Tesla wanted to change the world, and he did, thanks in part to his ceaseless promotion of a hyperconnected, efficient, and peaceful future world without messy wires and blown fuses. He showed people the future, sparks and all.

VI

POWER FROM THE EARTH AND SKY

TESLA'S WIRELESS WORLD SYSTEM

Tesla showed us that being a great entrepreneur—one who commercialized a critical standard that powers innovation 125 years later—isn't necessarily about the money. He acted as a quintessential engineer and humanist by finding ways to utilize the resources of the planet for the benefit of mankind.

—Gordana Vunjak-Novakovic,
Professor, Columbia University

In Tesla's tiny, darkened New York lab, guests gasped as Mark Twain held the inventor's vacuum lamp. The century was ending, heralded by this glowing orb in the hands of a steamboat captain-cum-literary genius—a romanticist of the past, skeptic of the present, and prophet of the future.

Tesla wanted to give the man who gave him comfort with *Tom Sawyer* as a lonely boy in the Balkans a piece of joy to hold in his hands. Yet as Twain looked down at the globe, phosphorescing due to an unseen force, it seemed as if he were holding a sullen cat that he couldn't pet. The man who had thrilled thousands by telling his wild mining-camp stories and adventures in King Arthur's court and by recounting real-life dinners given in his honor by the Emperor and Empress of Germany perhaps felt his mortality slipping

Tesla's 187-foot Wardenclyffe tower, shown in this 1904 photograph, represented the inventor's dream of transmitting wireless radio and power across the Atlantic.

The frontispiece of the 1916 edition of Mark Twain's *The Mysterious Stranger* shows a remote, mountainous, Eastern-European setting not unlike that of Tesla's childhood.

away. Twain's beloved daughter, Susy—never in robust health—would soon leave him, followed by his wife, Livy, a few years later. Perhaps Twain was wondering, as Tesla did, if the power that illuminated the lamp could reanimate human life. Was he musing over Mary Shelley's prospect that such a force could reshape humanity into something monstrous, or was he looking a restorative genie directly in the eye?

The great humorist's doleful expression straddled the border of humility and futility. In a literary attempt to acknowledge his friend's gift, Twain would later spin a tale titled *The Mysterious Stranger* of a boy growing up in the Balkans. The presumed homage, like so many of the writer's failed investments in inventions, was not well received, nor was it quite finished. Was it a dark, satiric depiction of Tesla being tempted by the devil or Twain's Faustian swan song? I'll leave that question to literary historians.

As an investor in new inventions, Twain must have been excited to be in Tesla's company. The writer loved novelty and technology, and despite repeated disappointments he always hoped to profit from the latest gizmo. Twain was jazzed when he saw Tesla's AC motor in action, proclaiming that it would "revolutionize the whole electric business of the world. It is the most valuable patent since the telephone."

In his role as a benevolent Wotan, Tesla entered the twentieth century thoroughly wedded to his idea that he could liberate the world from the darkness of isolation and war. His entire body had been permeated with electricity on many occasions, and the high voltage often caused temporary memory loss. In any case, a Manhattan building was no place to experiment with millions of volts of electricity. Having lost one lab to fire—caused, some Teslaphiles claim, by sabotage, although no one knows for sure—he needed an open space to try out some new ideas.

Tesla's marriage to his science was now, at the dawn of the twentieth century, fully consecrated, and his celibacy, at least according to his own vague self-description, ensured his priestly devotion to higher powers. Certainly Katherine Johnson, wife of the *Century Magazine* editor Robert Underwood Johnson, had deeply admired the tall, swarthy genius. Women's rights advocate and philanthropist Anne Morgan, the daughter of the imperious banker, was said to be "interested."

But Tesla swore to preserve all his strength and energy for the days to come; the élan vital of the inventor was the flame of the universe itself. The chimeric inventor could channel sparks from his fingertips, yet he needed the courage and wisdom to channel the energy of electricity toward endeavors that would affect civilization in profound ways.

Mountain Man

Based on the finding that he could make current jump short distances in his New York lab, Tesla was convinced he could send current much further, perhaps even around the world—if there was enough power at his disposal. Having reached an agreement with the small power utility El Paso Power in Colorado Springs, he headed for the Rockies in May 1899. With the support of Morgan and John Jacob Astor, the owner of the posh Waldorf Astoria hotel in which he resided in New York, Tesla assembled an enormous Tesla coil that became the bedrock for his experiments.

Back on the East Coast, General Electric was well on its way to creating the infrastructure for an *AC-based* national power system. Led by the competent manager Charles Coffin and the brilliant engineer Charles Proteus Steinmetz, Morgan's manufacturing conglomeration, sans Edison, was creating a new generation of ever-larger dynamos, transformers, turbines, and power transmission equipment. Like Tesla, Steinmetz had the tools and ability to create artificial lightning, although Steinmetz was now part of a well-funded corporate entity and Tesla was on his own.

What drew Tesla to the Rockies was the splendid isolation and the power above the peaks. In the summer, almost like clockwork, thunderstorms would coalesce over the mountains, creating crackling displays of lightning. Although he built his shed-like lab in the plains within sight of Pike's Peak, Tesla wanted to be close to the source of this natural electricity. Could it somehow be trapped and transported like his AC power? What happened to the current from the sky as it hit the earth? Was the earth an efficient transmitter of energy? If so, how did it work? Tesla brought these questions to the town of Colorado Springs, which had prospered from various mining operations deep in the mountains.

In the lightning that he was to create and send across the emptiness below the Front Range, Tesla saw freedom. It was a clean source of power that didn't have to be generated by burning coal, which created steam to turn the turbines of his dynamos. It was free energy that wouldn't be owned by anyone—not Morgan, not General Electric, not Westinghouse. It could be shared, broadcast like tiny seeds at the speed of light to any point of the world.

Tesla peeks out the door of his Colorado Springs shed-like laboratory in early summer, 1899.

Whether this was a selling point for J.P. Morgan, who initially had lent Tesla $150,000 in hopes that he would develop a "world telegraphy system," it's hard to make a case for Morgan's interest in creating world unity through universally accessible electricity. Tesla had promised the magnate he could send and receive signals to and from Europe. Morgan wanted to know stock results from the London Exchange and trade on the information before anyone else. He reached into his own pocket to finance Tesla and kept a wary eye on his daughter, Anne, who was said to be attracted to the handsome inventor.

Even though Marconi had been granted a patent on a wireless device—based on Tesla's designs—in 1896, Tesla promised Morgan that his system could transmit much more than information; it would be a power broadcaster as well. In the spring of 1899, he arrived in Colorado Springs, setting up his laboratory just outside of town at 6,000 feet. Accompanied by the wireless engineer Fritz Lowenstein, Tesla linked to the town's power grid and dramatically boosted the voltage "to make a number of discoveries of far-reaching importance."

Within a shed with a removable roof that Tesla hoped wouldn't catch fire was a transformer system connected to a ball-topped rod that could be raised and lowered. It looked like one of Franklin's giant lightning rods and was capable of hurling lightning bolts nearly one hundred feet—an astounding display at the time.

Tesla's Colorado research would yield four new patents for the wireless transmission of electricity. Although Tesla's technology was still far ahead of Marconi's, he lusted for more. The Serbian inventor wanted to build an entire system around air transmission of current, but also employ the earth as a conductor. His work with his lightning-producing coils and oscillators presaged the modern radio transmission era. More importantly, he discovered that when the current was in sync with that of the earth—at around six to eighteen cycles per second—a resonance occurred that made the transmission of electricity and mechanical energy surprisingly efficient.

So-called "standing or stationary" waves could link earth and sky, both of which contained plenty of energy. The earth, a mammoth magnet, creates heat from its molten core while the sun is constantly bombarding the earth and atmosphere with charged particles. An invisible, ionized shield around the earth called the ionosphere, extending 50 to 250 miles above sea level, protects us

The exterior of Tesla's Colorado Springs laboratory is shown against the backdrop of the Rocky Mountains, with the 142-foot telescoping "antenna" mast, topped by a thirty-inch copper-covered ball, extended upward from a hatch in the roof.

from most of the sun's potentially deadly radiation fusillade (which would later become a long-term health hazard for space travelers). Tesla realized that the energy from the sky was perpetual—it didn't need to be generated from water or coal—and could somehow be tapped for yet another source of free energy.

Tesla's ability to monitor the energy from the sky became so acute that he could predict the approach of oncoming storms with his equipment. One device detected a storm's energy even as the tempest was moving away from him. As Tesla recounted in his journal about his Colorado experiments:

> "It was thought that the lightning was now too far and it may have been about 50 miles away. All of a sudden the instrument began again to play, continuously increasing in strength, although the storm was moving away rapidly.... This was a wonderful and most interesting experience from the scientific point of view. It showed clearly the existence of *stationary waves.*"

In this image from Tesla's 1900 article "The Problem of Increasing Human Energy," his huge "magnifying transmitter" coil in Colorado Springs burns the nitrogen in the atmosphere, resulting in discharges of 12 million volts, measuring sixty-five feet across.

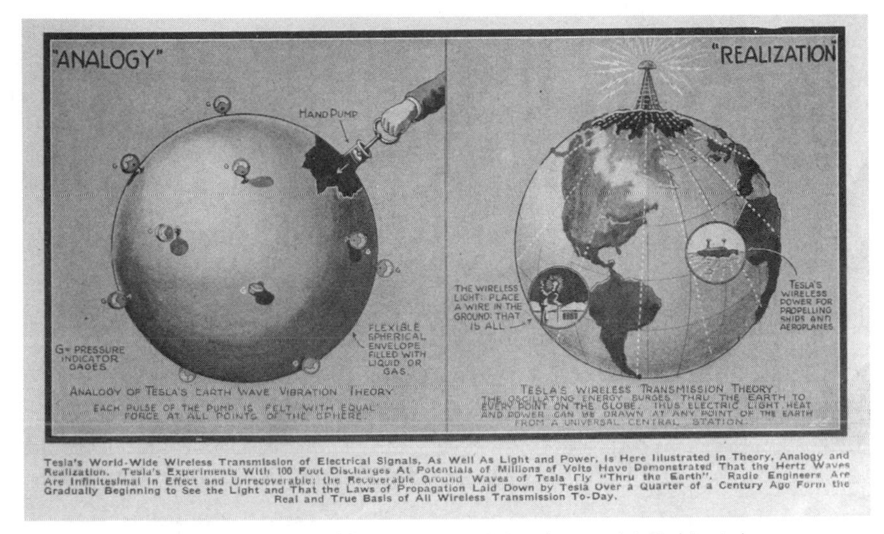

This illustration printed in the *Electrical Experimenter* in 1919 seeks to explain Tesla's wireless system theory. He planned to transmit electric power without wires using a network of magnifying transmitters that would send current though the earth to a "receiver" tuned to the resonant frequency of the transmitter. The left-hand drawing shows a mechanical analogy in which electric charge is a pneumatic fluid filling the earth, with the magnifying transmitter as a pump and the wireless receivers as pressure gauges. The right-hand drawing shows oscillating earth currents created by the magnifying transmitter tapped anywhere on the planet to power lightbulbs, vehicles, or aircraft.

If electricity could be channeled from the sky through the earth, Tesla reasoned that huge systems could link the planet, provided that the signals themselves were strong enough to overcome the resistance of the air, soil, and bedrock. With this supposition/discovery in hand, Tesla tore down his facility after eight months and headed back to New York.

Wardenclyffe

Tesla had trumpeted his Colorado Springs research in a long, illustrated piece titled "The Problem of Increasing Human Energy" in the June 1900 issue of *The Century*: With "absolute certitude, communication without wires to any point of the globe is practicable." In the same article, the inventor extends his thinking to "telephone and television." Yes, *television*, which was nearly a half-century away from practical use. Observing the numerous obstacles to public health and overall quality of life, he put forth an elaborate theory on the mass and energy of humanity and how to increase them by various means, including vegetarianism and the use of ozone for water purification.

At about the same time, Marconi had secured a $350,000 contract with Britain's Admiralty to install wireless sets on twenty-six ships along with six on-shore stations in the United Kingdom. The aggressive Marconi didn't want the British government to obtain a monopoly on the technology, so he and his team worked tirelessly to make the project successful.

Meanwhile, Tesla, residing at the Waldorf, was still trading off of his reputation, living in high style with access to all of the barons of finance and industry of the time. John Jacob Astor, who owned the Waldorf, had staked Tesla with $100,000. Tesla bought 200 acres on eastern Long Island, near Shoreham, to construct what he hoped would be a community of thousands of workers employed by him to transmit signals wirelessly.

Tesla's rival, Italian radio pioneer Guglielmo Marconi, poses with his early radio apparatus in 1896. The device at left, powered by high voltage from an induction coil (not shown), is the transmitter. The box at right is the receiver.

Tesla wanted not only to outdo Marconi—he had promised Morgan his system would offer multiple modes of communication—but also to prove that his "world telegraphy" was going to be global in scope. He wasn't talking about sending single letters of Morse code; he was assuring investors that he could link every continent with channels that could provide information *and* voice transmission, and he didn't stop there. As he recounted in his autobiography,

> "The greatest good will comes from technical improvements tending to unification and harmony, and my wireless transmitter is preeminently such. By its means the human voice and likeness will be reproduced everywhere and factories driven thousands of miles from waterfalls furnishing the power; aerial machines will be propelled around the earth without a stop and the sun's energy controlled to create lakes and rivers for motive purposes and transformation of arid deserts into fertile land."

Tesla's architect friend Stanford White, who had built the graceful arch that stands at the northern entrance of Washington Square in Greenwich Village, designed an elegant laboratory with neoclassic windows and a chimney capped by crown-like ironwork that was called a "wellhead" (its scrolled ironwork has since been restored). The building looked more like an estate garden house for an English lord than a working center for experimentation.

Behind the White building, Tesla planned to erect the Wardenclyffe tower, the heart of the complex. Powered by a 200-kilowatt generator on loan from Westinghouse, the pinnacle he envisioned would be the most powerful transmitter of any kind on earth at the dawn of the twentieth century. It would radiate waves of electromagnetic energy across the Atlantic and eventually be connected to a series of other stations across the globe, which would act as relays to boost intercontinental signals.

Tesla dreamed big at Wardenclyffe as Marconi and German engineers at Telefunken, which also had a facility on Long Island, labored on their wireless systems on a much less grandiose scale. Although Tesla originally called for a gargantuan 600-foot tower, he downscaled to 187 feet, the length carefully calibrated to push a signal across the Atlantic. Looking like a metallic

Rising above Stanford White's neoclassical laboratory design near Shoreham, New York, Tesla's tower was essentially a huge Tesla coil consisting of a wooden tower topped by a hemispherical sixty-eight-foot copper capacitive electrode.

and wooden mushroom on stilts, the tower was elegant in its own way, although Tesla ran out of money building it and repeatedly appealed to Morgan and his associates for more.

Less visible were tunnels running under and away from the tower. One central shaft, accessed by a spiral staircase, went down some 120 feet to ground the tower's transmitter in the bedrock. While it's not precisely clear why the tunnels were needed, later observers speculated they were needed for drainage since the water table in the area was high (around 80 feet). As the work progressed—and the money ran out—Tesla continued to ship his lunches in from the kitchens of the Waldorf by train. Asking Morgan for $200,000 to finish the project and bring it online, Tesla became desperate to prove that it could transmit a signal at a great distance.

Tesla didn't stop with the world system, though. He proposed a device that would be powered by it, presaging today's wildly popular pocket appliances:

> “A cheap and simple device, which might be carried in one's pocket, may then be set up somewhere on sea and land and it will record the world's news or such special messages as may be intended for it.”

Tesla's system, anticipating as it did cellphones, email, geographic positioning, and even texting, was probably too much for Morgan to handle; the banker was mainly interested in getting stock quotes and collecting art.

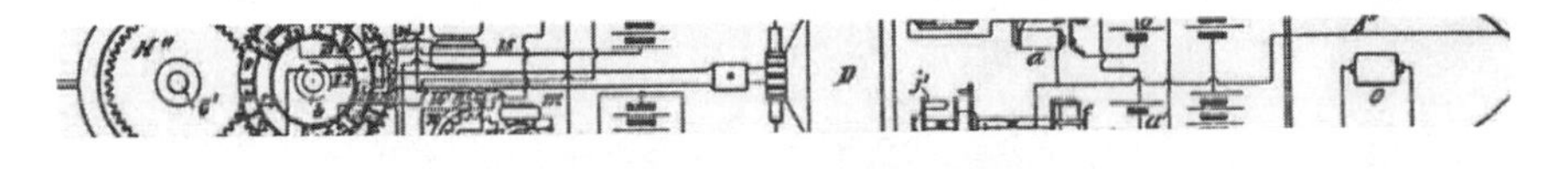

TESLA'S PROPHECIES

In a 1915 essay in *Manufacturer's Record,* Tesla challenged the world with even loftier possibilities: "The impossible has happened, the wildest dreams have been surpassed and the astounded world is asking: What is coming next?"

The inventor would answer with an array of innovations that didn't materialize until decades later:

- **Cogeneration.** Tesla didn't call it that, but it was a way of turning heat generated by manufacturing or other processes into electricity. "In the very near future, such (thermal) waste will be looked upon as criminal." He called for a system of "thermodynamic transformers."
- **Hydroelectric Power.** Then, as now, water power was a clean, efficient source of energy. Nothing had to be burned to produce power, and nature supplied the kinetic energy, 85 percent of which could be converted into electricity. At the time, Tesla saw expanded hydropower replacing the annual combustion of 120 million tons of coal. This was well before the "big dam" era of the 1930s through President Franklin Roosevelt's rural electrification programs, which lit up the American South and West.
- **Electric Propulsion.** Tesla embraced the possibility of universal electricity powering all forms of transportation—wirelessly. As the use of the steam engine faded, the electric motor slowly replaced it on a large scale. Most railroad locomotives today, for example, are electric motors driven by diesel-powered dynamos. Everything from cars to buses can run on electricity. As early as 1904, Tesla had proposed the use of electric motors in autos. "We would advise you that we consider the induction motor poorly adapted for this class of services," Westinghouse replied. Of course, in light of the half million orders for the new Tesla Model 3 electric compact sedan (as of this writing) this would seem like a poor decision on Westinghouse's part.
- **Disinfection.** Although he wasn't a biologist, Tesla knew that electrical current could sterilize just about anything. He predicted electrical devices such as "smoke annihilators, dust absorbers, ozonizers and sterilizers," all of which are in use today and are increasingly prevalent in large, industrial systems. Tesla had even patented a device to create ozone.

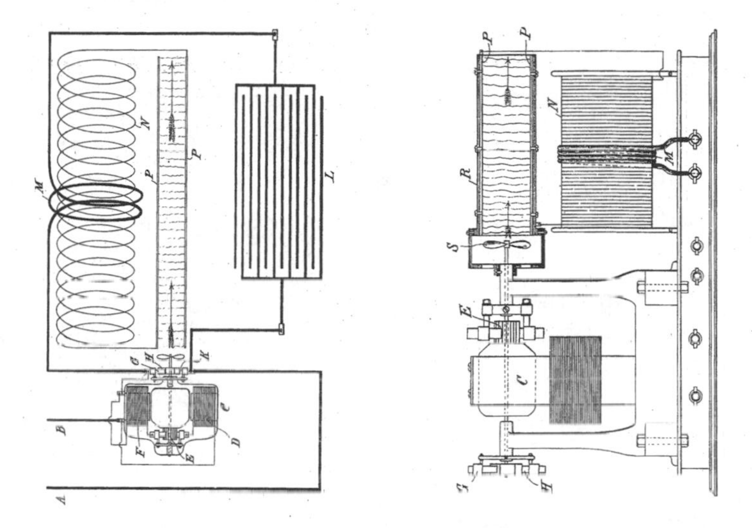

Diagrams from Tesla's 1896 patent "Apparatus for Producing Ozone."

- **Navigation.** In addition to powering ships with electric motors through wireless networks, Tesla called for a global navigation system "that will enable vessels to get at any time accurate bearings and other practical data." Yes, Tesla was referring to global positioning systems (GPS), a giveaway on most smartphones today, although this was long before the age of geosynchronous satellites.

- **Wireless Weapons.** While World War I saw the widespread use of horrible weapons such as poison gas, machine guns, torpedoes, mines, mammoth projectile guns, and "buzz bombs," Tesla knew that his wireless technology would spawn several new generations of weapons. "I believe the teleautomatic aerial torpedo will make the large siege gun, on which so much dependence is placed at present, obsolete." His remote, electronically controlled weapons emerged decades later in the form of "smart bombs," drones, and ballistic and cruise missiles.

- **Solar Energy.** Even in the throes of financial and professional turmoil, Tesla was considering the long-term effects on the world of "dirty energy" that would increasingly rely upon the consumption of fossil fuels. Disparaging the combustion of fuel as "barbarous and wantonly wasteful," Tesla turned his mind to the sun and its bountiful gift of widely dispersed energy. With back-of-the-envelope math, he then concluded that about "100,000 horsepower per square mile" could be converted to electricity, although the scale of such energy recovery was "beyond the pale of practical." While Tesla did not know how to reap power from the sun on a large scale, he ended this stream of thought with a hope that wireless energy "to any distance" would be the great equalizer, so that "humanity will be united, wars will be made impossible and peace will reign supreme."

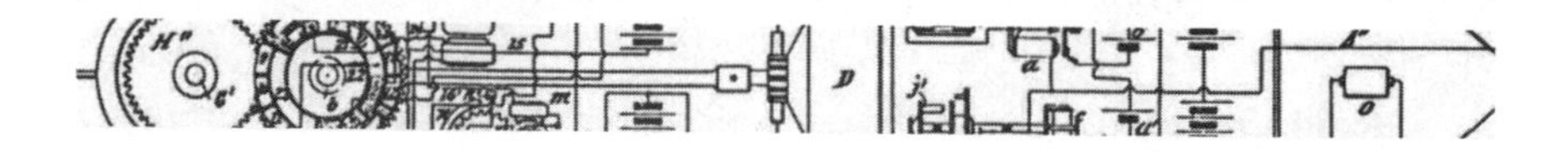

A New View of Time and Space

By 1915—the same year that Tesla published his visions for the future in *Manufacturer's Record*—the scientific world was moving at a breakneck pace as Albert Einstein published his general theory of relativity.

As Tesla was contemplating the future of energy—having even devised an equation ($E = mV^2/2$) to model total human energy (E), as a function of the mass of humanity (m) and the velocity of human change (V)—the young Swiss patent clerk–turned–theoretical physicist was focusing his genius on visualizing his place in the universe. Like Tesla, Albert Einstein saw things in his head and had a similar ability to walk himself through a dynamic mindscape in his "thought experiments."

Taking the stage at the prestigious Prussian Academy of Sciences in Berlin, Einstein presented his theories to show how time, space, matter,

This 1920 photograph shows Albert Einstein in his office at the University of Berlin.

gravity, and energy were connected in a new world that could be explained by a series of formulas that showed how the universe worked.

A decade earlier Einstein had formulated his special theory of relativity, which held that nothing could travel faster than the speed of light, at which point matter would turn into energy. Even more perplexing to Einstein was a question that vexed Newton and Galileo: How did gravity work as a force in the universe? The physicist's conclusion was that it warped space-time like a flexible fabric. Gravity *bent* space and impacted time itself. Waves of this force rippled the universe like a pebble hitting a tranquil pond. Using his powerful visualization capacity, Einstein *saw* that light itself could be curved by the weak force of gravity as it traveled through space.

In the first decades of the twentieth century, Einstein became as famous as Tesla had become in the late 1890s, morphing into an international rock star who changed everything we knew about the mechanics of the universe. Thanks to Einstein, modern astronomers can detect the presence of distant planets in other galaxies, among many other scientific advances.

Does gravity travel the same way as Tesla's waves? Can it be harnessed to propel us beyond our planet into other galaxies without the use of rocket fuel? As an inventor, Tesla's core pursuit was practical application, not abstract theory. He certainly thought that wireless energy could propel electric ships, and he publicly disagreed with Einstein's theories, speculating that some kinds of energy might move faster than the speed of light. Regardless of whether the speed of light is nature's true upper limit, though, some new discoveries on Einstein's "gravitational waves," which were finally observed in late 2015, might open up new inquiries into how to send energy through the earth and perhaps the cosmos—making Tesla's telegeodynamics concept a reality.

Possibly Tesla was too disruptive for his time, but he never lost sight of his obligation to humanity as an inventor:

> " Only thru annihilation of distance in every respect, as the conveyance of intelligence, transport of passengers and supplies and transmission of energy will conditions be brought about some day, insuring permanency of friendly relations. What we now want most is closer

> contact and better understanding between individuals and communities all over the earth, and the elimination of that fanatic devotion to exalted ideals of national egoism and pride which is always prone to plunge the world into primeval barbarism and strife.... The only remedy is a system immune against interruption. It has been perfected, it exists, and all that is necessary is to put it in operation. ”

Ultimately, Tesla saw himself as a humanitarian above all else and believed he could actually unify humanity and end all wars through the implementation of his "World System." A century after this bold assertion, we are still a violent lot, but history has made much of his technical vision a reality.

In 1919, Tesla made the following reflection:

> “Not a few technical men, very able in their special departments, but dominated by a pedantic spirit and nearsighted, have asserted that excepting the induction motor I have given to the world little of practical use. This is a grievous mistake. A new idea must not be judged by its immediate results.”

For Tesla, one invention—even one entire transmission system—wasn't enough. His dreams were bigger and kept him going for the rest of his life. Like Einstein seeking a grand unification theory, Tesla wanted to create a global superhighway for electricity as well as communications.

Why look at the big picture when so many of us are profusely sweating the small stuff? Because we need to! Look at what's staring at us like some predatory beast: global warming, terrorism, the problem of feeding billions. Of course, no one person can conquer these mammoth problems alone, but if more of us make a habit of looking at long-term scenarios and solutions—perhaps one application or invention at a time—we can do a lot of good for humanity. Just look at the kind of impact the Google search engine has had on our lives, thanks in large part to the "Teslaphile" Larry Page!

Tesla had a guiding vision in the most idealistic sense when creating his magnifying transmitter and transformer: He wanted peace for humanity through technology. As he noted in his autobiography,

> “We all must have an ideal to govern our conduct and insure contentment, but it is immaterial whether it be one of creed, art, science or anything else, so long as it fulfills the function of a dematerializing force. It is essential to the peaceful existence of humanity as a whole that one common conception should prevail.”

Having an expanded and common worldview may well be necessary for our survival—to prevent us from killing each other off, if nothing else—yet it is constantly challenged by a world focused on short-term results. Nearly every company is vexed by societal and market pressures to show improved results every quarter. This "short-termism" spills over into other areas such as education and culture, and it can be toxic when trying to attack broad-based problems, which is why we still need the ambitious, unruly ideas that came from Tesla and people who think like him.

Thinking big starts with big questions:

- What kind of impact do you hope your idea or creation has on the world at large? What long-term impact would it have?
- What legacy do you hope to leave after you die? In order to look back on your life without regrets, what do you need to accomplish?

Vol. V. Whole No. 53 September, 1917 Number

U. S. Blows Up Tesla Radio Tower

SUSPECTING that German spies were using the big wireless tower erected at Shoreham, L. I., about twenty years ago by Nikola Tesla, the Federal Government ordered the tower destroyed and it was recently demolished with dynamite. During the past month several strangers had been seen lurking about the place.

Tesla erected the tower, which was about 185 feet high, with a well about 100 feet deep, for use in experimenting with the transmission of electrical energy for power and lighting purposes by wireless. The equipment cost nearly $200,000.

The late J. P. Morgan backed Nikola Tesla with the money to build this remarkable steel tower, that he might experiment in wireless even before people knew of Marconi. A complete description, revised by Dr. Tesla himself, of this unique and ultra-powerful radio plant was given in the March, 1916, issue of THE ELECTRICAL EXPERIMENTER. Everyone interested in the study of high frequency currents should not fail to study that discourse as it contains the theory of how this master electrician proposed to charge this lofty antenna with thousands of kilowatts of high frequency electrical energy, then to radiate it thru the earth and run ships, factories and street cars with "wireless power."

Most of our readers have, no doubt, read about the famous Tesla wireless tower, which structure involved the expenditure of a vast sum of money and engineering talent. From this lofty structure, which was designed some 20 years ago by Dr. Tesla and his associates, there was to be propagated an electric wave of such intensity that it could charge the earth to such a potential that the effect of the wave or charge could be felt in the utmost confines of the globe.

Further, it may be said that Tesla, all in all, does not believe in the modern Hertzian wave theory of wireless transmission at all. Several other engineers of note have also gone on record as stating their belief to be in accordance with Dr. Tesla's. More wonderful still is the fact that this scientist pro-

Two Views of the Last Minutes of Tesla's Gigantic Radio Tower at Shoreham, L. I., New York, As It Was Being Demolished by the Federal Government. It Was Suspected That German Spies Were Using the Tower for Radio-Communication Purposes. It Stood 185 Feet Above the Ground and Cost About $200,000. Tesla Had Not Used It For Several Years.
Photos by American Press Association

mulgated his basic theory of *earth current* transmission a great many years ago in some of his patents and other publications. Briefly explained, the Tesla theory is that a wireless tower, such as that here illustrated and specially constructed to have a high capacity, acts as a huge electric condenser. This is charged by a suitable high frequency, high voltage apparatus and a current is discharged into the earth periodically and in the form of a high frequency alternating wave. The electric wave is then supposed to travel thru the earth along its surface shell and in turn to manifest its presence at any point where there might be erected a similar high capacity tower to that above described.

A simple analogy to this action is the following: Take a hollow spherical chamber filled with a liquid, such as water; and then, at two diametrically opposite points, let us place, respectively, a small piston pump, such as a bicycle pump, and an indicator, such as a pressure gage. Now, if we suck some of the water into the pump and force back into the ball by pushing on the pisto handle, this change in pressure will be i dicated on the ga secured to the o posite side of th sphere. In th way the Tesla ear currents are su posed to act.

The patents Dr. Tesla are b sically quite diffe ent from those Marconi and othe in the wirele telegraphic fiel In the nature things this wou be expected to the case, as Tes believes and h designed apparat intended for t *transmission* *large amounts* *electrical energ* while the energ received in t transmission of i telligence wireles ly amounts to b a few millionths an ampere in mo cases by the tin the current transmitted h been picked up thousand mil away. In the Her zian wave syste as it has been explained and believed in, t energy is transmitted with a very large lo to the receptor by electro-magnetic wav which pass out laterally from the transmi ting wire into space. In Tesla's system t energy radiated is not used, but the curre is led to earth and to an elevated termin while the energy is transmitted by a proce of *conduction*. That is, the earth receives large number of powerful high frequen electric shocks every second, and these a the same as the pump piston in the analog

Quoting from one of Tesla's early pa ents on this point: "It is to be noted th the phenomenon here involved in the tran mission of electrical energy is one of *tr conduction* and is not to be confound with the phenomena of *electrical radiatio* which have heretofore been observed, a which, from the very nature and mode propagation, would render practically in possible the transmission of any apprecia amount of energy to such distances as a of practical importance."

VII

WEATHERING THE STORM

THE FALL OF WARDENCLYFFE

Tesla is one of the few scientists who dedicated himself to achieving peace in a very tangible way, by developing inventions that can prevent wars. What he offered mankind today unfortunately has been forgotten to a certain extent—the ideal of a scientist who feels and knows that true scientific progress can never, and should not, be dislocated from the good and the beautiful.

—Aleksandar Protic, Director of the Tesla Memory Project, UNESCO

The latticed Wardenclyffe tower careened to one side like a ship listing to starboard after the dynamite blast did its work. It was a wounded thing of beauty, nothing like the streamlined needles of today or messy cellphone towers with punkish transmission caps. It was Stanford White's abstract form of a mushroom cloud with a sturdy concrete base. But now it was a harpooned leviathan, reduced to a blubbered pile of ugly scrap to be sold for a fraction of what it cost to build.

Although the Long Island facility was never fully functional, the U.S. government at the time had seized all possible radio transmission facilities as the United States entered World War I and cut off any possibility of

This *Electrical Experimenter* article from September 1917 displays dramatic snapshots of Tesla's radio tower listing to one side during its demolition, which began in July of that year.

spying through the wireless. The demolition of Tesla's tower in 1917 was a more simple matter, though: After a series of misfortunes, in 1915 he was forced to sign over his Wardenclyffe deed to George Boldt, the manager of the Waldorf, to settle a $20,000 debt. After the hotelier's death in 1916, his estate gave the order: Sacrifice the tower to scrap dealers. More than a dozen years after the century had begun with such promise, not only was Europe convulsing into a mechanized killing field, but everything that Tesla had built and owned was now in the hands of somebody else.

A Financial Sea Change

Long before the last section of the Wardenclyffe tower crashed to the ground, history had rudely intervened in Tesla's plans. The robber baron era of personal financing of technology began to unravel quickly when President William McKinley was assassinated on September 6, 1901. Gone was a business-friendly White House that embraced everything the trusts and corporate interests hoped to accomplish with their monopolies. Teddy Roosevelt became president, beginning a new age of progressive trust-busting and hostility to what he called "malefactors of great wealth." Undaunted by the change in the political scenery, Tesla pressed on, urging White to make his tower bigger still. "My calculations show that with such a structure I could reach across the Pacific," Tesla wrote White on September 13, a week after McKinley's assassination.

For Morgan, Tesla was headed in the wrong direction. Marconi's devices were much cheaper to build and operate, and the Italian was more than willing to charge people for the use of his system—a sharp contrast to Tesla's humanitarian purpose of providing universal power. Historians of technology claim that Tesla's quest at Wardenclyffe imploded when Marconi successfully transmitted a single letter—"s"—across the Atlantic on December 6, 1901. At that point, whatever political and financial capital Tesla could garner evaporated. It was if the stock price in Nikola Tesla Inc. had crashed. Tesla was now an also-ran, and the entire New York finance community saw Marconi as *the* leader in wireless telegraphy.

This photograph shows the towers of Marconi's Poldhu Wireless Station in Cornwall, England, which sent the first transatlantic radio signals in December 1901, effectively putting an end to Tesla's Wardenclyffe dream.

Even though Tesla fired up the tower for one trial, which neighbors testified had lit up the sky like a "corona" for miles, he lost the race to prove that radio—a technology that he had invented—worked over great distances. Tesla would graciously acknowledge Morgan's "noble generosity" in a 1904 piece in *Electrical World and Engineer,* but the banker was pulling back on his commitment to Tesla.

Lacking Insull's and GE's robust stock capitalization, the inventor had only had a virtual handshake agreement with Morgan and his other investors. After Marconi started to garner capital and attention from all over the world, Tesla had to scrape for every dollar as Wardenclyffe continued to devour money. To add insult to injury, the U.S. patent office, succumbing to Marconi's growing international celebrity, reversed itself and awarded the patent for radio technology to the Italian inventor. This action would later

256 February 6, 1904 ELECTRICAL WORLD AND ENGINEER. Vol. XLIII, No. 6.

The Annual Dinner of the American Institute of Electrical Engineers.

The annual dinner of the American Institute of Electrical Engineers will be held in the main ball room of the Waldorf-Astoria, New York City, Thursday, February 11, 1904, at 7 P.M., precisely. The guest of honor will be Mr. Thomas Alva Edison, and it is expected that an opportunity will be given to meet Mr. Edison at an informal reception, at half-past 6. This dinner will commemorate the twenty-fifth anniversary of the introduction of the incandescent lamp, and will also celebrate Mr. Edison's birthday. Upon this occasion will also be presented the deed of gift of the Edison Medal Association, which has raised a fund of several thousand dollars.

The following guests have accepted invitations: Mr. Ambrose Swasey, president American Society of Mechanical Engineers; Dr. A. R. Ledoux, president American Institute of Mining Engineers; Col. Robert Clowry, president Western Union Telegraph Company; Mr. George G. Ward, vice-president and manager Commercial Cable Company; Mr. W. H. Baker, vice-president Postal Telegraph Cable Company; Mr. John Fritz, founder Bethlehem Steel & Iron Works; Mr. W. H. Fletcher, president Engineers' Club; Mr. A. B. Chandler, president Postal Telegraph-Cable Company.

Autographed souvenir menus have been prepared containing a colored photogravure of Mr. Edison and two original poems prepared by Mr. R. R. Bowker. The exercises will be as follows: Salutatory address by President B. J. Arnold; presentation of the medal fund and deed of gift by Mr. Samuel Insull, chairman of the Edison Medal Association; acceptance of same on behalf of the Institute by Dr. A. E. Kennelly, of Harvard University, past president; address on behalf of the colleges and universities by Prof. Cyrus F. Brackett, of Princeton University; address on behalf of the Association of Edison Illuminating Companies by President J. B. McCall; address on behalf of the National Electric Light Association by President Charles L. Edgar.

Mr. Edison has flatly declined to speak, but in response to the toast in his honor has agreed to send from the table a telegraphic acknowledgment. It is a great many years since he used the key. He will use one of the original quad sets built by him for, and loaned by the Western Union Telegraph Company. The message will be received in the banquet hall on a Postal quad of latest date by President A. B. Chandler, of the Postal Telegraph Cable Company, and will then be read to the audience. These arrangements are in the hands of Mr. C. P. Bruch, assistant general manager of the Postal system, and Mr. J. C. Barclay, chief engineer of the Western Union system. A number of special cable messages and telegrams will also be received at the same time.

The seating arrangements provide for the accommodation of eight persons at each table. Orders for seats should be sent in at once and should be accompanied by cash or check, payable to Mr. Ralph W. Pope, secretary. Price of tickets, without wine: Gentlemen, $7; ladies, $5; admission to galleries, $1. In order to secure accommodation, responses should reach Mr. Pope not later than February 9, 1904, 95 Liberty Street, New York. Mr. Arthur Williams, chairman of the committee on decorations, has made elaborate preparations for ornamenting and illuminating the ball room, and the effect will be very pretty and appropriate.

A Striking Tesla Manifesto.

We reproduce herewith in slightly reduced fac-simile the first page of a four-page circular which has been issued this week by Mr. Nikola Tesla in a large square envelope bearing a large red wax seal with the initials, "N. T." At the back of the page which we reproduce is given a list of 93 patents issued in this country to Mr. Tesla. The fourth page is blank. The third page has a little vignette of Niagara Falls and is devoted to quotations from various utterances of Mr. Tesla. The first of these is from his lecture delivered

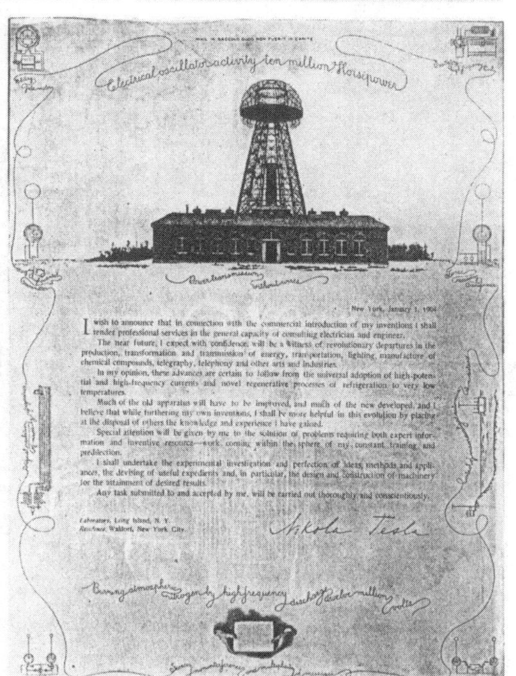

New York, January 1, 1904

I wish to announce that in connection with the commercial introduction of my inventions I shall tender professional services in the general capacity of consulting electrician and engineer.

The near future, I expect with confidence, will be a witness of revolutionary departures in the production, transformation and transmission of energy, transportation, lighting, manufacture of chemical compounds, telegraphy, telephony and other arts and industries.

In my opinion, these advances are certain to follow from the universal adoption of high-potential and high-frequency currents and novel regenerative processes of refrigeration to very low temperatures.

Much of the old apparatus will have to be improved, and much of the new developed, and I believe that while furthering my own inventions, I shall be more helpful in this evolution by placing at the disposal of others the knowledge and experience I have gained.

Special attention will be given by me to the solution of problems requiring both expert information and inventive resource—work coming within the sphere of my constant training and predilection.

I shall undertake the experimental investigation and perfection of ideas, methods and appliances, the devising of useful expedients and, in particular, the design and construction of machinery for the attainment of desired results.

Any task submitted to and accepted by me, will be carried out thoroughly and conscientiously.

Laboratory, Long Island, N. Y.
Residence, Waldorf, New York City.

Nikola Tesla

PAGE FROM CIRCULAR SHOWING TESLA TOWER, WARDENCLIFFE, LONG ISLAND.

in 1893 before the Franklin Institute and the National Electric Light Association, as to transmission of intelligible signals and power to any distance without the use of wires. The second quotation is from his article on the problem of increasing human energy, which appeared in the *Century Magazine* in June, 1900, dealing with virtually the same subject. The third item quotes from his patents, Nos. 645,576 and 649,621, dealing with the transmission of electrical energy in any quantity to any distance, with transmitting and receiving apparatus movable as in ships or balloons. The circular is an extremely interesting one. It is most sumptuously got up on vellum paper and altogether constitutes a manifesto worthy of the original genius issuing it. It is to be gathered from the circular that Mr. Tesla proposes to enter the field of consulting engineership, in which he already has enjoyed an extensive connection here and abroad.

Proudly displaying his Wardenclyffe tower and diagrams from several of his patents, Tesla's manifesto, which was reproduced in a 1904 edition of *Electrical World and Engineer,* offers the inventor's services as a consulting engineer.

trigger a patent suit by Tesla, although the case wasn't resolved in his favor until after his death.

By 1903, Tesla was effectively running out of cash. Without capital to pay his workers, he would have to cease operations at Wardenclyffe, yet he was unrelenting in his optimism for his grand scheme. The following year, still hopeful that the cash flow would resume, he brushed off the problem in a trade magazine piece as due to "unforeseen delays" that were a "blessing in disguise." However, in a disastrous chain reaction, a series of misfortunes continued to derail Tesla, who solicited capital from Morgan's circle and pledged royalties from his patents to the banker. In 1904, what remained of his Colorado Springs lab was torn down and sold for the value of its lumber. His friend and architect Stanford White, who had endured numerous redesigns and upgrades on the Wardenclyffe tower at little or no compensation, was shot and killed

New York American

PRICE ONE CENT

HARRY THAW KILLS STANFORD WHITE ON ROOF GARDEN!

GAS TRUST ACCEPTS 80-CENT RATE.

Consolidated's Chief Lawyer Admits Defeat in The American's Fight.

400,000 PERSONS AFFECTED

Legal Rate Will Be Accepted from Responsible Men Till the Courts Decide.

VICTORY BY INJUNCTIONS

Attorney Shearn Gives All the Credit to Mr. Hearst and Makes a Final Move.

Keyport girl whose body, with that of her sweetheart, was found in Raritan Bay yesterday.

DON'T HAVE TO PAY A TRUST IN MISSOURI

Patron Is Upheld by Court in Claim That Corporation Restrains Trade.

GIRL SLAIN IN ROWBOAT BY HER SUITOR.

Battled with the Frenzied Youth and Screamed in Vain for Help.

HIS SUICIDE FOLLOWED

Fellow Townsmen Draw Mantle of Charity Over Case, Calling It "Accident.

FIVE ICE TRUST HEADS SENT TO JAIL FOR YEAR.

For First Time in History of the Nation Monopoly Chiefs Are Thus Punished.

ACTION TAKEN IN TOLEDO

Conspiracy and Extortionate Prices the Charges Which Brought Convictions.

STANFORD WHITE

Who was shot to death last night while seated at a table in Madison Square Roof Garden.

Originator of "23, Skidoo for Yours," Brought to Justice

SHOOTS ARCHITECT IN BACK AS HE SITS TALKING TO WOMAN

Slayer, Captured, Gives Police False Name, but Is Immediately Recognized.

A TRAGIC OPENING NIGHT

Singing Girls Pluckily Keep Up the Music as Shots Ring Out—Prisoner Refuses to Say a Word in His Cell.

Excitement Blurs the Facts.

On June 26, 1906, the *New York American* broke the sensational news of Stanford White's murder by Harry Thaw, the jealous husband of showgirl Evelyn Nesbit. Wardenclyffe, which had been mostly designed by White, was unfinished and underfunded at the time of his death.

in June 1906 while attending the opening of a show at his Madison Square Garden stadium at 26th and Madison Avenue. Apparently his killer was the jealous husband of a young showgirl with whom White reportedly had sexual relations (a fictionalized version of the sensational drama was later depicted in the 1955 film *The Girl in the Red Velvet Swing*). White's loss was painful for Tesla, who had first met him through Robert and Katherine Johnson's circle.

Nevertheless, Tesla, forging on, championed his discovery of stationary waves in Colorado, implying they could be used for communication, even as rapid advances in wireless and telephone systems proceeded. Tesla envisioned Wardenclyffe as a future complex for telecommunication, a nexus for what later become not one, but two colossal industries.

Of course, Tesla wasn't alone in anticipating the creation of two new industrial giants in the Bell telephone network and what would become the Radio Corporation of America. Still a visionary, he believed that what he had begun at Wardenclyffe would eclipse them both. The realization that he was running out of money and not getting the adulation and support he had in the past from the New York elite triggered a nervous breakdown in 1905.

In the interim, Tesla would sell his royalty rights in the equipment he developed for Westinghouse for just over $200,000. Those revenues were at least $12 million at the time. The move literally cost Tesla billions, as he would have received royalties on every Westinghouse AC motor sold.

The punishing blow to Tesla's prestige, income, and ego was the awarding of the Nobel Prize to Marconi in 1909 for his invention of the radio—an event that stung Tesla deeply, since the Italian's technology was clearly based on Tesla's schematics and oscillator principles. (There's some debate as to how Marconi acquired Tesla's ideas and schematics before the turn of the century, but as a 1943 court ruling pointed out, the basic principles of radio transmission came from Tesla's patent on a four-circuit electrical system, which he had applied for in 1897 and which was granted to him in 1900.)

Since Tesla badly needed money, in 1906 he started promoting a bladeless turbine, which was tested in Waterside Station, New York, over the years 1911 and 1912. Revolutionary as they were, the turbines were not universally adopted and became a footnote in Tesla's career. Moreover, they didn't generate the cash he was seeking, and Tesla was forced to vacate the

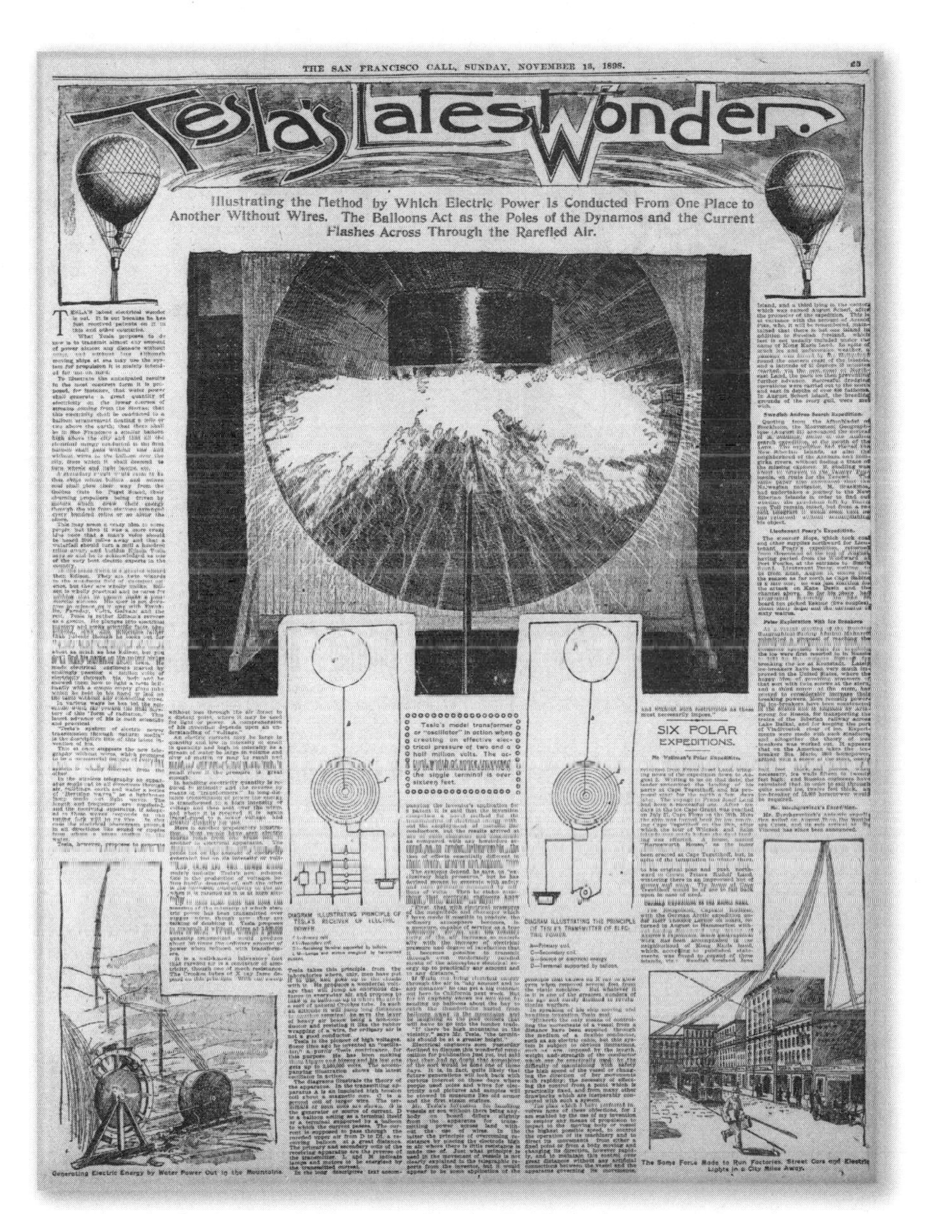

THE SAN FRANCISCO CALL, SUNDAY, NOVEMBER 13, 1898.

Tesla's Latest Wonder

Illustrating the Method by Which Electric Power Is Conducted From One Place to Another Without Wires. The Balloons Act as the Poles of the Dynamos and the Current Flashes Across Through the Rarefied Air.

Tesla's model transformer or "oscillator" in action when creating an effective electrical pressure of two and a half million volts. The ac- [illegible] the single terminal is over sixteen feet.

SIX POLAR EXPEDITIONS.

Diagram Illustrating Principle of Tesla's Receiver of Electric Power.

Diagram Illustrating the Principle of Tesla's Transmitter of Electric Power.

Generating Electric Energy by Water Power Out in the Mountains.

The Same Force Made to Run Factories, Street Cars and Electric Lights in a City Miles Away.

An article in the *San Francisco Call* from 1898 describes Tesla's plan for transmitting wireless power with balloons, featuring diagrams he included in his four-circuit electrical system patent. In 1943, a court ruled that Tesla had developed the basic principles of radio transmission before Marconi.

Tesla demonstrates an electrical apparatus in his office in 1916. Despite numerous disappointments during the first decades of the twentieth century, the inventor continued to develop, promote, and improve his technologies.

Waldorf after his friend and financial backer John Jacob Astor perished in the *Titanic* disaster on April 15, 1912. The "unsinkable" ship, heralded as the most invulnerable vessel of its day, had telegraphed its Mayday signal with one of Marconi's wireless transmitters.

Years after Wardenclyffe went dark, Tesla resumed his campaign for funding from the House of Morgan, which had been headed by Morgan's son, Jack, since J.P.'s death in 1913. Tesla's offices, now in the Woolworth Building, still had the Wardenclyffe tower on his letterhead. Tesla hoped that the younger Morgan wouldn't blackball him as he father had, even though Tesla was unable to repay the father's original $150,000 loan. Setting his sights low, Tesla asked the younger Morgan for $10,000, even though in April 1913 Jack Morgan made it fairly clear that "it was impossible for him to become interested in further inventions," meaning funding Wardenclyffe. The younger Morgan did, however, loan Tesla $20,000 in four installments

Bonus feature

Learn more about Nikola Tesla with this interactive content.

1. Download and launch the Nikola Tesla app.

2. Position your mobile device over the images on the following pages.
Allow the scanning process to identify them.
Note: Keep the page flat and provide adequate lighting.

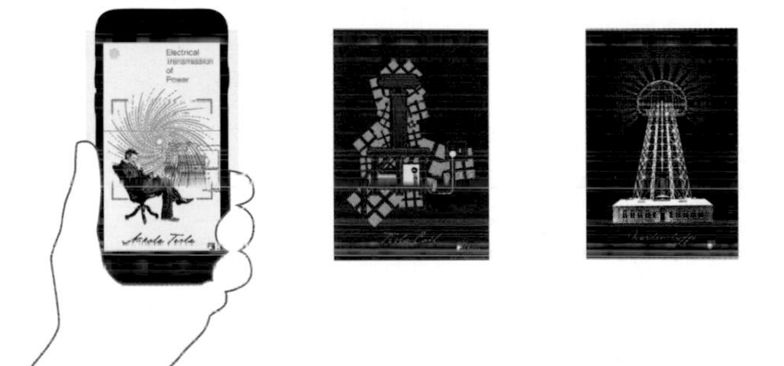

3. Unlock the magic of augmented reality!™

Interactive content produced by Yetzer Studio

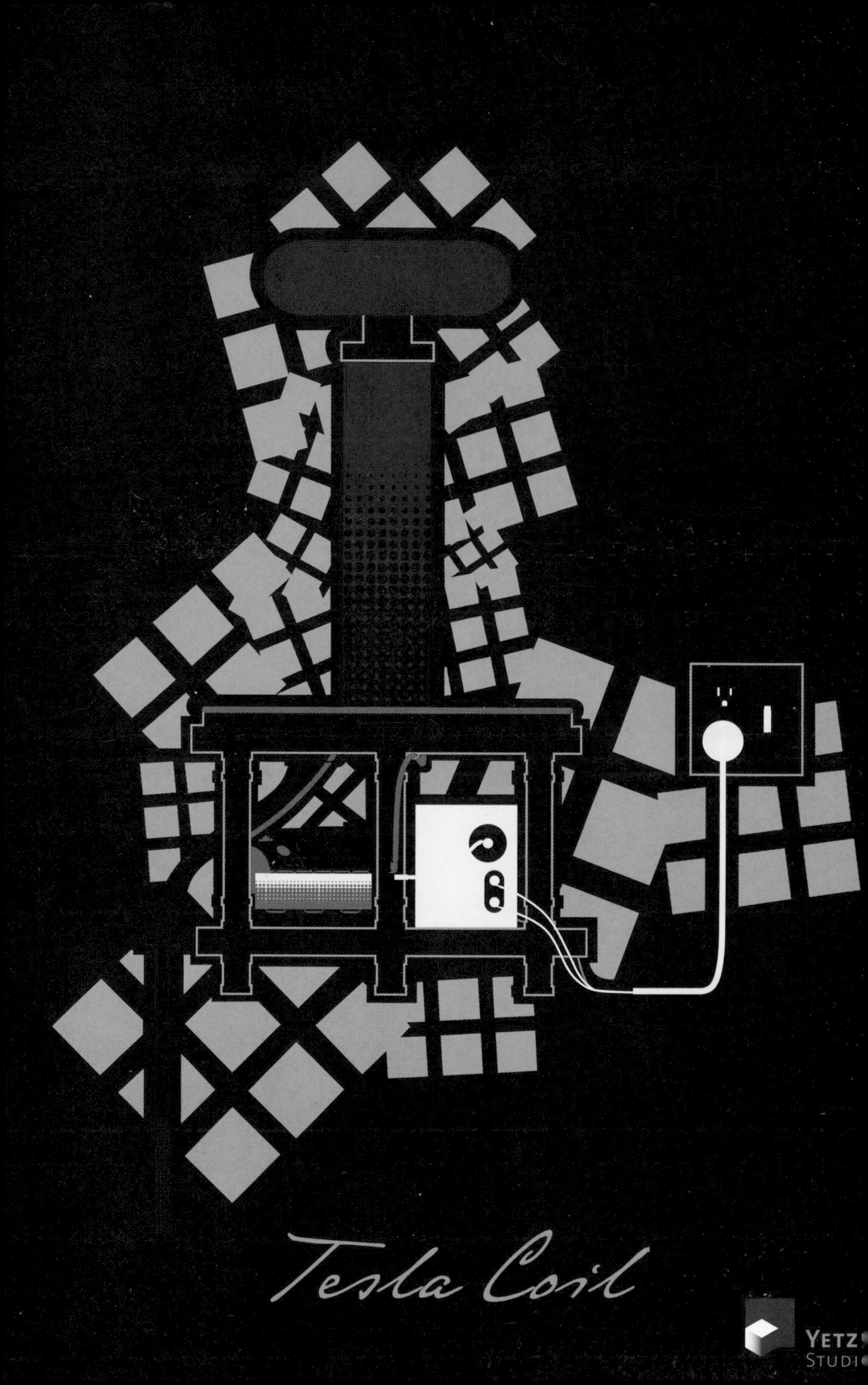
Tesla Coil

Electrical
Transmission
of
Power
INVENTOR
Nikola Tesla
Nikola Tesla

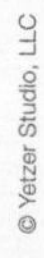
© Yetzer Studio, LLC

YETZER
STUDIO

Wardenclyffe
YETZ
STUDI

after his father died, but only for development of his bladeless turbine. With World War I demanding credit from the world's leading investment banking house, Tesla's further requests ended up in Jack's out basket.

Admitting that Wardenclyffe had cost him more than $500,000, Tesla again courted Morgan Jr. in July 1914 to extend more credit, based on future income from Tesla's turbine:

> “Even if I should never be able to carry out what I have undertaken, I still can acquit myself of your father's generous loan provided that certain steps, which involve the expenditure of a relatively small amount of money, are promptly taken.”

Earlier in the year, Jack Morgan had stiffed Tesla, having his secretary write him, “Mr. Morgan is not prepared to make any further advances of money to you.” By the beginning of 1915, Morgan was emphatically cutting Tesla off, noting Tesla would have to find financing “without any further assistance from me.” Westinghouse, who had died the previous year, didn't direct his company to further finance Tesla either, although at least the lines of communication were open between Tesla and Westinghouse's successors.

Tesla's cruel financial decline could have been braked at the end of that year when it was announced that Tesla and Edison would share the Nobel Prize in physics. The award, of course, would have boosted Tesla's prestige immensely and probably solved his money woes. But the story had been misreported. Neither man would be going to Stockholm.

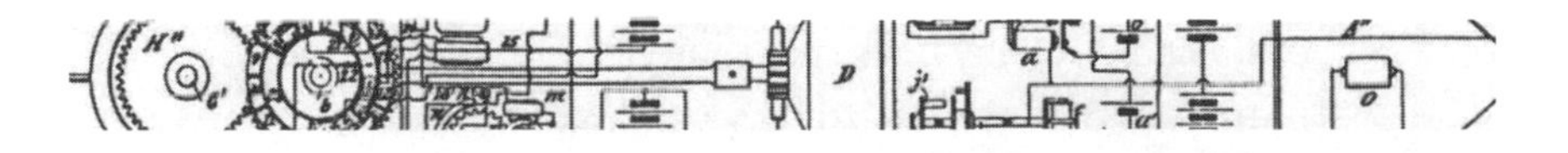

TESLA'S INDOMITABLE SPIRIT

Possibly Tesla's most enduring trait was his resilience: He fused belief with dogged determination, even when the fates dictated otherwise. As a boy, he escaped danger and demise on numerous occasions:

> “I was almost drowned a dozen times; was nearly boiled alive and just mist being cremated. I was entombed, lost and frozen.... But as I recall these incidents to my mind I feel convinced that my preservation was not altogether accidental.”

As discussed in chapter III, some instances of survival were attributable to his unusual visions, although even those wouldn't have saved him if he hadn't possessed a very strong will to survive in the first place.

Tesla also had several nervous breakdowns, mostly from overwork and exhaustion, beginning in his student days. As shown in his description of a breakdown he suffered after dropping out of college, these were true physical collapses:

> “My ear was thus over thirteen times more sensitive. Yet at that time I was, so to speak, stone deaf in comparison with the acuteness of my hearing while under the nervous strain. In Budapest I could hear the ticking of a watch with three rooms between me and the time-piece. A fly alighting on a table in the room would cause a dull thud in my ear.... I had to support my bed on rubber cushions to get any rest at all. The roaring noises from near and far often produced the effect of spoken words.... The sun's rays, when periodically intercepted, would cause blows of such force on my brain that they would stun me.... My pulse varied from a few to two hundred and sixty beats and all the tissues of the body quivered with twitchings and tremors which was perhaps the hardest to bear.”

With the moral support of his friend Anthony Szigeti, who forced him to go on long, outdoor walks with him, Tesla was able to recover, noting that his recovery became "a sacred vow, a question of life and death. I knew that I would perish if I failed."

Tesla always found a way to get back to work after a collapse. Of course, his kind of labor was both a narcotic and medicine for him, so he had to learn to take care of his mind and body by acknowledging their limits:

> **“When I am all but used up I simply...naturally fall asleep .... [I]f I attempt to continue the interrupted train of thought I feel a veritable mental nausea.”**

Tesla also found great virtue in routine. Following the 1895 fire that destroyed years of his work and plunged him into a profound depression, he consciously strove to get himself on a regular schedule to prevent the extremes of overwork and breakdown. Although most of us have mixed feelings about the daily nine-to-five, if our work is filled with purpose, it can be healing, as it often was for Tesla. As he noted in his autobiography, he saw a payoff in steady work:

> **“My belief is firm in a law of compensation. The true rewards are ever in proportion to the labor and sacrifices made.”**

Tesla remained focused on his ultimate goal. Belief in what he was doing always got him past his epic highs and lows. Once he got out of the valley, he knew he could get to the next peak. Part of this belief in why you should climb the mountain after the dark days in the valley is having a larger purpose, which Tesla did. Through friendships, self-care, routine, obligation to humanity, and sheer will he was able to overcome numerous brushes with death, nervous breakdowns, failures, and financial setbacks, refocusing and hewing to his narrative, which became even more strident as the world headed into mass poverty, fascism, and war.

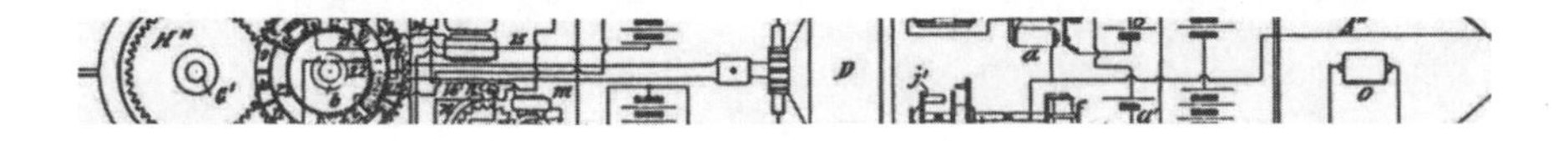

The following information is required by the Board of Examiners. Failure to give details will delay action on the application.

1. Give full name, date and place of birth.
2. Give general and technical education, where and how acquired.
3. State under which clause or clauses, a, b, c, d, of Section 4, Article II, of the Constitution, application is made, and give full record of professional career, with particular reference to the period of responsible charge or the experience upon which the application is based. clauses a and d.

Dates here	APPLICANT'S RECORD
1857	Born Smiljan, Lika, border county of Austria-H
1873	Graduated Higher Realschule, Carlstadt, Croatia
1873-7	Polytechnic School Gratz, - mathematics and physi
1877-9	Univ. of Prague, Bohemia, Philosophical
	Degrees: M.A. Yale 1894; LLD Columbia Univ.
	D.Sc. Vienna Polytechnic.
1881	Began practical career at Budapest, Hungary, where was made his first electrical invent a telephone repeater, and where was conceived the idea of the rotating magnetic field; later enga in various branches of engineering in Fran and Germany.
1884	Came to the United States, of which he is a naturalized citizen
1886	Invented system of arc lighting
1888	Invented the Tesla motor and system of alternating-current transmission, two-pha and three-phase.
1889	Invented system of electrical conversion and distribution by oscillatory discharges
1890	Invented generator of high-frequency curren
1891	Invented system of transmitting energy over a single wire without return. Invented the Tesla coil or transformer
1891-3	Investigated high-frequency effects and phenomena.

A snapshot of Tesla's 1916 application for the AIEE fellowship shows a detailed record of his accomplishments, all written down in his own hand.

Dates here	Applicant's Record (continued)
1893	Invented system of wireless transmission of intelligence
1894-5	Invented mechanical oscillators and generators of electrical oscillations
1896-8	Researches and discoveries in radiations material streams and emanations
1897	Invented high-potential magnifying transmitter
1897-1905	Invention and development of a system of transmitting energy without wires
1898.	Invention of a system of transmitting energy with minimum loss by refrigeration.
1901-2	Invention of system for magnifying feeble effects
To present—	Most important recent work discovery of a new mechanical principle which has been embodied in a variety of machines such as reversible gas and steam turbines, pumps, blowers, air compressors, etc, having a greatly increased output per unit weight as compared with any other form of machine for like service.

(signed) Nikola Tesla

NOTE: The Applicant's personal signature, in ink, must appear at the end of this record.

If not sufficient space, the record may be continued on separate sheets of this size.

Failure and Fortitude

No one is immune from hardship. While I was studying what Tesla did after Wardenclyffe was abandoned and he went broke, I, too, endured some stormy times. My Freedom of Information Act requests to various branches of the government weren't going anywhere. All I had was the FBI file telling me that the agency had shipped Tesla's papers to some unnamed government warehouse, à la *Raiders of the Lost Ark*.

About the time I hit this research roadblock in 2008, the stock market collapsed. Lehman Brothers became the largest business bankruptcy in history, credit dried up, and a panic ensued. The government had to bail out the banks, mortgage lenders like Freddie Mac and Fannie Mae, and the giant insurance company AIG. The ensuing recession claimed millions of jobs, shut down the U.S. housing market, and forced a new suite of regulations of the financial industry.

Sitting at my home-based Bloomberg financial data terminal, where I had been writing a column about investing for the news service for seven prosperous years, I could see the world coming apart. Nobody would be paying much attention to a personal column about investing because everyone, including the folks on Wall Street who indirectly subsidized my writing, would be in high survival mode. The following year, I was cut off from my livelihood when Bloomberg dropped my column in the wake of losing major clients like Bear Stearns and Lehman Brothers.

In that horrible season, I lost my mother, who had encouraged me to write and create, after a long battle with leukemia. I also lost a dear friend to cancer. Grief was like a stray dog that followed me everywhere. Then, my wife Katherine was diagnosed with breast cancer and had to undergo surgery and treatment. She bravely endured aggressive chemo sessions, searing radiation, and two trips to intensive care—all while taking classes at a community college, and acing her courses! During this period, even the heavens were jolting us: Our home was literally struck by lightning—twice.

Just as Tesla found it necessary to develop a daily routine to prevent overexertion, we concentrated on getting my wife healthy again. We modified our diets and threw out personal care products that contained carcinogenic chemicals and endocrine disrupters such as parabens.

I was also forced to reevaluate my role as an independent columnist and writer. By necessity, I redefined my professional raison d'être: I was no longer just a journalist, but an *advocate* for my readers. I prepared to explore difficult, undesirable corners of my beat, such as aging, long-term care, investor protection and financial abuse, college financing and debt, and the dark side of retirement. Gradually, I found new outlets for my writing, which led to more speaking engagements and special investigative projects through groups like the Nation Institute Investigative Fund. I started a blog with *Forbes* that eventually garnered up to a half-million monthly views. After a few more bumps in the road, I was getting assignments from AARP, the *New York Times,* and other national publications.

During this period, Tesla was like a dove that flew in my window. I was heartened by how the inventor revisited and reengaged his skills in designing small, practical mechanical devices like the speedometer and bladeless turbine. There was a hidden reserve to Tesla's indomitable spirit. In the face of financial ruin, he still kept his visionary focus and kept himself in the public eye by speaking and writing about the ideas he cared about. While his big-vision project would occupy him for the rest of his life, he never gave up, even with a major war approaching and the world moving at great speed into a perilous century.

From the end of the Wardenclyffe era to his death in 1943, Tesla warmed to "the greatness of our age" by redefining himself, probing deeper for new ideas as the evolving twentieth century presented even more horrific challenges.

As Tesla reflected on the failure of his Wardenclyffe project, he had some sage words:

> “My project was retarded by laws of nature. The world was not prepared for it. It was too far ahead of its time. But the same laws will prevail in the end and make it a triumphal success.”

Resilience always requires that you put personal failures into perspective, placing them against the much larger backdrop of history. That's a revelation that came to me in the pits of my despair. It wasn't about me; it was about my role in helping my family, which meant that my writing and other work were about broader service and not *self*-service.

In my own life, I faced some draconian setbacks with illness in my family and loss of income, so, like Tesla, I had to redefine my goals and put one foot in front of the other until I was able to climb out of the chasm in which I found myself. The cliché "Don't sweat the small stuff" is probably good advice most of the time, but when you're trying to recover from a health or career crisis that shakes up your worldview and faith in the future, the small stuff can be wonderfully distracting! So get off Facebook and the medication with nasty side effects. Improve your diet. Take a walk every day.

Failure doesn't have to be a one-way descent into the abyss. In fact, it can be powerfully instructive. Your creative powers don't automatically go dark because you've had a few setbacks. In fact, your creative drive may become even stronger. Although his dream project failed to attract the funding it needed, Tesla moved on after Wardenclyffe, applying his creative skills to smaller problems, where he had some success. He took long walks, wrote, and read the classics. He kept his visions alive by writing about them and eventually courting journalists who still cared about his idealistic views. Although he was effectively out of business in the risky world of big experimental projects, he never stopped being creative.

Persistence of vision is also important. What if Steve Jobs, having been bounced from Apple Inc. in the 1990s, had decided that he was done with the technology business? We never would have had the iPhone® or the iPad® or the hundreds of innovations that came with them. You don't have to put your life on hold for your dream, either. You can still tinker in your garage, as many Silicon Valley entrepreneurs did—and are still doing. You can follow your passion at night because there's still nothing wrong with paying your bills.

Although Tesla spent an enormous amount of time in solitude, he wouldn't have been able to overcome every setback had he not had friends and associates with whom to commiserate. Don't forget that you need to talk (and listen) with friends, family, and possible business associates while you're recovering from a major setback or just slogging through an enormous project. I'm sure all my friends and acquaintances were sick of me telling them I was "still working" on my Tesla book. "Haven't you already written that?" they would groan. But still, it's extremely important to be out in the world during good and bad times. Show your work. Network. Attend lectures. Volunteer in the community. Be with your family. Be curious. Talk to people who have done something important.

There is no one script for getting through the tough times, and whatever route you take may be untidy and nonlinear. Diversify your experience by seeing and doing different things. Go see a dance performance or sculpture exhibit. Bang on a drum, play piano, or go bowling. Take a walk in the woods on a snowy winter day. Be *present* in the life outside yourself and your goal.

Resilience is the life energy that gets you up on those dark, winter mornings. If, like Tesla and so many others, you believe that life energy or the soul itself is literally like electrical current—which we can transmit—we can also *receive* free energy from the universe. Perhaps there is a feedback loop that supplies the energy we need (and opportunities we crave) to succeed if we just hold on long enough. Grit can keep you going.

VIII

MAN FOR ALL SEASONS

ADAPTING TO A NEW, TURBULENT CENTURY

Tesla was one of the few who have bequeathed our planet with not only messages of peace and inventions of technological progress, but also with the frequency of peace on which we should act...

—Mirjana Prljevic, Executive Director,
Peace and Crises Management Foundation

The squid-like machine resembled something out of a novel by H.G. Wells or Jules Verne, but it was *so* twentieth century in 1903. Steam punkers would have been thrilled to stage an electronic rave around this leviathan, a cylindrical unit about three stories high with two levels of oval portholes in the side. It looked more like a conning tower from Captain Nemo's sub than what it was: an engineering revolution called a turbogenerator. Nevertheless, Sam Insull, now running the Chicago Edison Company on Tesla's technology, was ready to fire up the unwieldy beast.

GE's best engineers had designed this imposing device—the largest steam turbine in the world at the time—on speculation, and it was going to be spectacular if it worked. If it failed, Insull surmised that he would lose all his investment (insurance wouldn't cover things like this). But if it succeeded, he could supply Chicago and other cities with Tesla's AC power.

This illustration from Burnham and Bennett's *Plan of Chicago* (1909) shows the "proposed boulevard to connect the north and south sides of the river." The plan laid the foundation for a modern, efficient urban metropolis powered by Tesla's AC current.

The gargantuan 5,000-kilowatt turbogenerator, designed by GE and installed in 1903 at the Commonwealth Electric's Fisk Street Generating Station in Chicago.

Planning for the future was Chicago's forte at the time. Six years later, Daniel Burnham and Edward Bennett would put forth a comprehensive master plan for the city meant to accommodate a rapidly growing population. Chicago was not only reinventing itself with a central plan that incorporated generous green space, lakefront parks, and industrial zones; it was reengineering itself with massive power plants that would provide electricity to steel mills, auto plants, and meatpacking houses. The city's leading minds were engaged in mega-engineering guaranteed to improve quality of life. Water from Lake Michigan could be filtered and disinfected, and the effluent from the city sent down the Chicago River by reversing its flow (sorry, St. Louis, Memphis, and New Orleans). Horses and steam engines came off the streets. The air, ground, and water cleared up. A city that had been built on swamps gave birth to the skyscraper, the elevator, and the modern factory and suburban home.

Long since departed from Edison and the hated Morgan cartel, Insull was financing his own electrical empire and bringing AC power to Chicago, the Midwest, and beyond. After consolidating several smaller companies, Insull wanted to go big and supply power to an entire major city that was then the fastest-growing metropolis in the world. Tesla was there in spirit in the cavernous space on Fisk Street, a new coal-fired power plant that hugged the Chicago River on the city's South Side.

Although Insull originally considered putting antiquated reciprocating steam engines in the plant, he rejected their installation since they were at their operating limits. They *could* generate the kind of juice he needed to power the city's electrified trolley system, which Insull had just won a contract to operate. But in order to generate enough power for his needs, the engines would have had to occupy an area equal to Chicago's entire downtown. That wouldn't work.

The old engines with giant, flaying arms generated mechanical energy from pulsating steam, which drove leather pulleys hooked up to dynamos. The system was incredibly inefficient and prone to numerous breakdowns. Too much energy was lost between the motors and the generators.

Insull personally took on the financial risk when he asked GE to design a turbogenerator, yet it was a masterpiece of design, vanquishing the nineteenth-century engines by channeling the steam directly from the boiler to drive the turbines, which shared a common, vertical shaft with the dynamo. Even more appealing to Insull was that his new 5,000-kilowatt unit occupied one-tenth the space of the reciprocal units and cost one-third less.

At the crucial moment when the switch was to be thrown to start the turbogenerator, Insull's top engineer, Fred Sargent, suggested that Insull leave the building in case the unit killed everyone in its vicinity.

"Why?" Insull asked.

"This is a dangerous business," Sargent replied. *"The damned thing might blow up!"*

"Well," Insull said with the matter-of-fact resignation that came of having staked everything on this one machine. "If it blows up, I'll blow up with it anyway. I'll stay."

The generator clanked and chortled, but it eventually worked. Insull moved on to build ever-larger central-station plants, selling electricity to the masses.

The Second Industrial Revolution, powered by Tesla's operating system, was making everything from mass-produced goods to public transportation affordable and ubiquitous. Now every city could have streetlights and electrically powered public transit. You could live in the suburbs and easily get into a city's central business district via an electric commuter train. You could enjoy an array of appliances in your home. You could not only safely illuminate every room, but also cool them with electric fans. You could vacuum the floor instead of beating a rug. Radios would become the most popular entertainment in the coming decades until the age of television. Cities in hot climates, like Miami, New Orleans, Atlanta, and Phoenix, could have air conditioning, which was invented by Willis Carrier in 1902.

Tesla had a hand in it all.

Back at Wardenclyffe, though, Tesla had seen better days. The government stepped in to put his patent suit against Marconi on hiatus as all forms of electronic devices were corralled by government edict. Although Woodrow Wilson didn't want to join the "war to end all wars," the sinking of Britain's RMS *Lusitania* by a German U-boat two years earlier had started the drumbeat for war. Taking the lives of nearly 1,200 people, the cruise ship was carrying some war material, although its primary purpose was civilian transport from New York to Liverpool. Tesla had briefly pitched the idea of using his tower as an electronic "shield" against enemy invasion, but Boldt and the government didn't want to take any chances.

With regret, Tesla left the luxury of the Waldorf in 1917, never again calling it his permanent residence. After a farewell dinner at the Johnsons', at which he was dressed like an earl with his "cane, white gloves, and his favorite green suede high-tops," writes Marc Seifer, Tesla left for Chicago in an attempt to develop his bladeless turbine for the Pyle National Corporation. Katherine Johnson, though suffering with the flu, was dressed in a florid gown to say goodbye to him with tears in her eyes, heartbroken at the sight of the inventor leaving her home, her city, and her life.

Back to Square One

Taking walks from his Chicago hotel, which was a few blocks from his triumph at the World's Columbian Exposition two decades removed, Tesla mused over what might have been had Wardenclyffe transmitted that first transatlantic signal instead of Marconi or had he obtained the capital he needed to provide universal power and information across entire oceans. He stared curiously at the triple green domes of what had been the fair's grand Palace of Fine Arts building, guarded by caryatids—the giant Greek maidens acting as columns. It was as if the goddesses of antiquity were holding up this temple of pure idealism, although it had gone to seed after the fair. It's not known if Tesla ever met with Julius Rosenwald, the Sears Roebuck executive and philanthropist who rescued the Beaux Arts masterpiece and

This photograph shows the original main entrance for the Palace of Fine Arts, with its magnificent Greek caryatids, at the World Columbian Exposition in 1893. It became the home of Chicago's popular Museum of Science and Industry (see page 104) in 1933.

converted it into the Museum of Science and Industry nine years later. The museum would be modeled after a similar, though smaller museum in Germany dedicated to science and technology.

Nor is it precisely known if Tesla walked a few more blocks to see what was transpiring at the young University of Chicago, a progressive institution exploring everything from the nature of the atom to the reformation of American education. Within a year, the university would appoint Albert Michelson, a German-born U.S. navy officer, as the head of its nascent physics department. Michelson's invention of the interferometer allowed scientists to measure the diameter of stars, among other things, and he had won the Nobel Prize in Physics in 1907 for his precision optical instruments. In 1887, the famous Michelson-Morley experiment had determined the speed of light.

Throughout most of 1918, as the war wound down, Tesla was in Chicago working on his new turbine, which was hampered by numerous technical issues. Although he said he enjoyed working with the Pyle engineers, the company refused to cover his expenses, which he estimated at more than $12,000. When the Pyle company insulted him by sending him a check for $1,500, Tesla, although deep in debt, sent the check back and left town. In the interim, his New York office was seized, and his patents expired. Although he was making some money from his improvement on the automobile speedometer, which he developed with the Waltham Watch Company and patented in 1918, it was hardly enough to cover his bills. He also developed valves and a "fluid diode" or "logic gate" that, according to engineer Leland Anderson, could be used in "logic circuits and simple fluid computers." However, after trips to Milwaukee and Boston to work on the turbine and speedometer, Tesla wasn't able to negotiate enough in advances and royalties to cover his growing costs.

In 1919, Tesla's admirer and promoter Hugo Gernsback, publisher of *The Electrical Experimenter*, convinced him to serialize his "autobiography" in his magazine. In relatively short vignettes, Tesla recounted the anguish of his brother's death, his education, his years working for Edison, and his early discoveries. Tesla's articles were a hit, propelling the magazine's circulation above 100,000, which made it the *Popular Science* of its day. Eventually

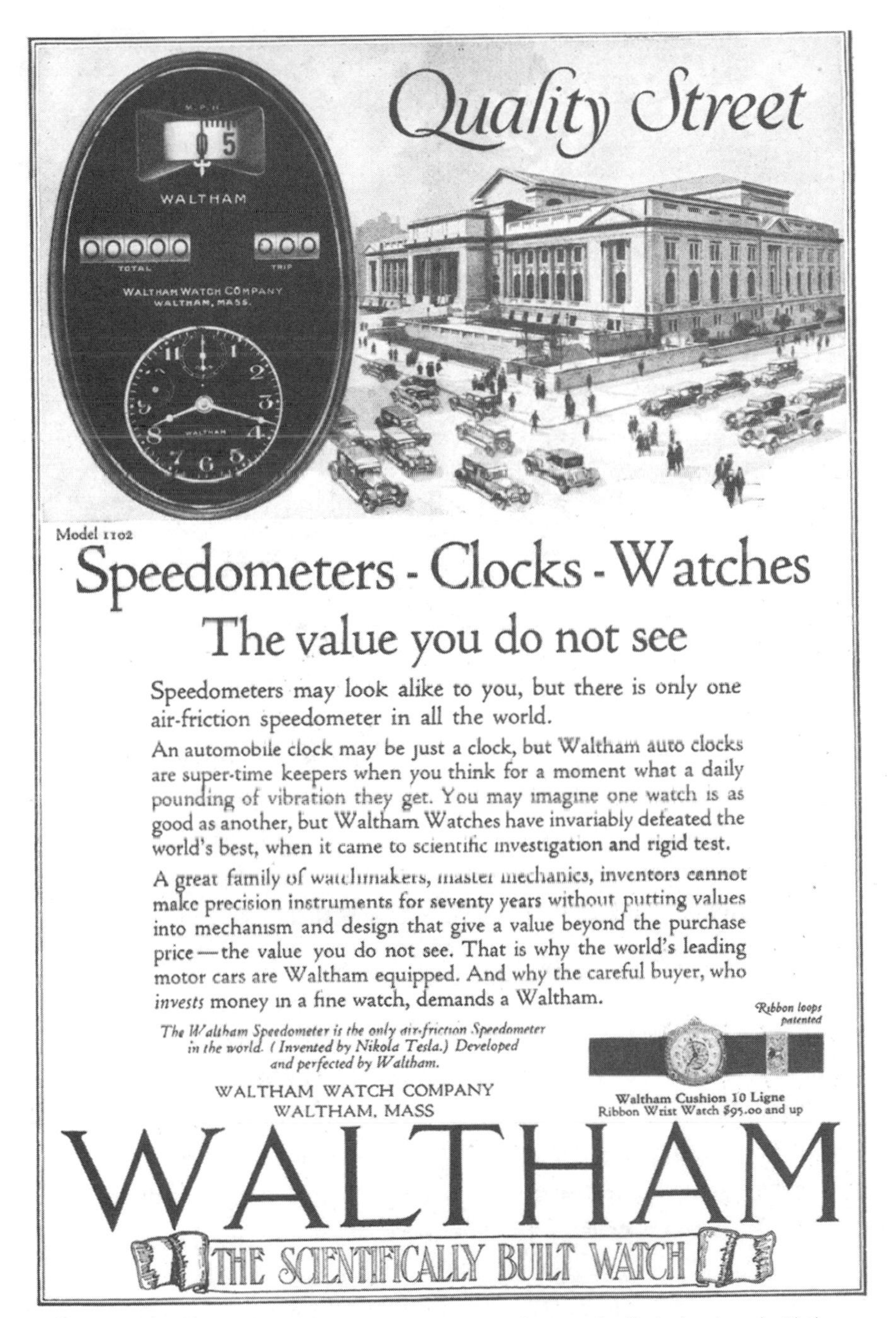

A 1922 advertisement for the automobile air-friction speedometer that Tesla developed with the Waltham Watch Company in 1908.

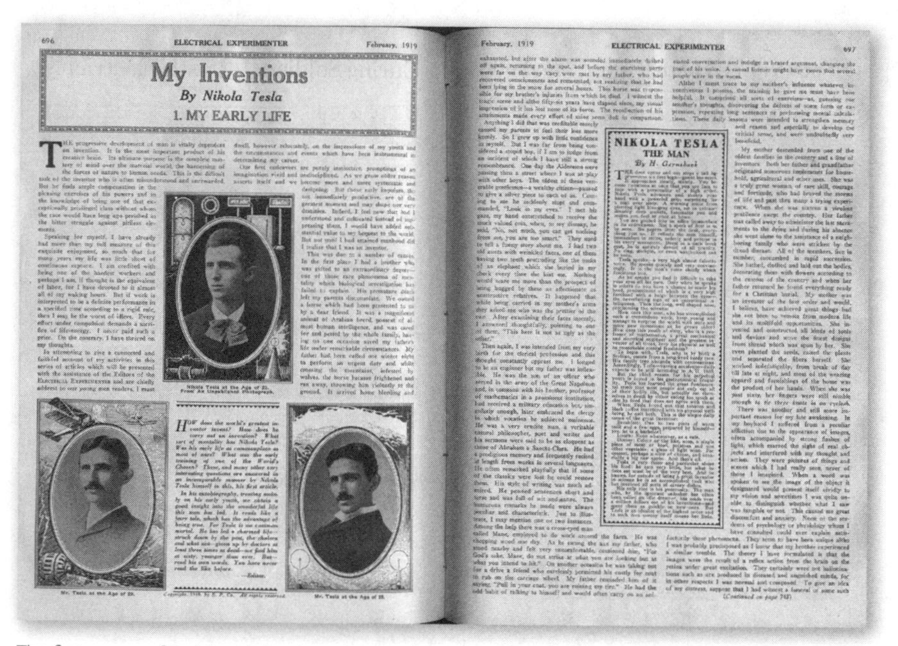
696 ELECTRICAL EXPERIMENTER February, 1919

My Inventions

By Nikola Tesla

1. MY EARLY LIFE

February, 1919 ELECTRICAL EXPERIMENTER 697

NIKOLA TESLA

THE MAN

By H. Gernsback

The first pages of Tesla's serialized autobiography in *The Electrical Experimenter,* which propelled the magazine's circulation above 100,000.

collected and titled *My Inventions*, the autobiography (quoted copiously throughout this book) provided some revelations about the enigmatic inventor. He was gracious in his depiction of Morgan, Edison, and others who alternately helped him and made his life difficult. He offered insights into his spirituality and waxed poetic on his new bladeless turbine, which he felt could revolutionize energy transmission:

> “But the prospective effect of the rotating field was not to render worthless existing machinery; on the contrary, it was to give it additional value.... My turbine is an advance of a character entirely different. It is a radical departure in the sense that its success would mean the abandonment of the antiquated types of prime movers on which billions of dollars have been spent.”

In the Roaring Twenties, Tesla's reputation rose somewhat, although his business affairs were a train wreck and his persona turned inward. Back in

Tesla considered his bladeless turbine, shown here with the upper half of casing removed, a truly paradigm-shifting technology that would make all modern heat-engines obsolete. As Yale electrical engineering professor Charles Scott quipped, "That would make quite a pile of scrap."

New York, he settled into the Hotel St. Regis. Always reluctant to shake hands and picky about what he consumed, he was now germ phobic and a strict vegetarian. He began roaming the city at night and would feed the pigeons near the library at 42nd Street, circle the block three times before entering his hotel, and avoid sidewalk cracks. Without a spouse or partner and separated from the Johnsons—Robert had been appointed ambassador to Italy by President Wilson—Tesla became a specter of his joyful self.

One of Tesla's last attempts to revive his Wardenclyffe dream was directed to the Westinghouse Electric and Manufacturing Company, then headed by E.M. Herr. In a letter to Herr in October 1920, Tesla sold the virtues of his "World System":

> "As you can see from the papers, the General Electric Company is evidently endeavoring to realize my "World System," but they will never succeed with any such apparatus.... If you are desirous to inaugurate a wireless system a century ahead of that in use at present, I can put you in a position to do so."

Since Tesla had torn up his royalty contract with Westinghouse years earlier, the electrical giant had no financial obligation to deal with him and had gone into the radio business, having pioneered KDKA, one of the country's first radio stations, in Pittsburgh. The station went on the air on November 2 of that year. Within ten days of KDKA going live, Tesla was badgering Herr to honor an earlier "conviction" that "nothing would be turned down that [Tesla] may put before the company."

But Herr wasn't interested in Tesla's ideas. He was shepherding a system of the company's own design that worked and would become a commercial success as the world grew to adopt and love radio as a major form of mass media. Echoing the bitterness that he must have felt over others developing and profiting from his ideas, Tesla maligned the systems in use to Herr:

> “The whole world is now committed to ridiculously inefficient and expensive installations which will all have to be torn down and reconstructed as soon as my system is put in operation.”

Concluding with “the wireless transmission of energy is my life work,” Tesla implored Herr to arrange a meeting with him. Not more than three days later, though, Herr replied, “I cannot proceed further with any development of your activities in this line.”

Tesla's last connection with the Westinghouse Company would resume fourteen years later when it agreed to contract with him as a “consulting engineer” for $125 a month—a generous amount, considering it was in the middle of the Great Depression—but a depressing coda for a man who had made the company billions.

Tesla Repositions His Business

Despite his dead-end with Westinghouse, Tesla continued to refine his speedometer with the Waltham Company. The inventor had repositioned himself as an all-purpose engineering inventor in his new offices on West 40th Street, not far from the future site of the Empire State Building. Tesla's letterhead from that time boasted of “steam and gas turbines; blowers, compressors, vacuum pumps; fountains; mechanical oscillators; precision instruments; high-frequency dynamos; lightning protectors; interference preventers; oscillation transformers and scientific novelties.”

Tesla doesn't specify what he means by “scientific novelties,” but it could have been an oblique reference to his remote-control boat, an electric fountain, or telegeodynamics. There was no mention of Wardenclyffe; the tower logo was replaced by a cutaway illustration of an uninspired bladeless turbine.

With his new brand—no longer was he promoting himself as the world-

NIKOLA TESLA COMPANY

8 West 40th St.
TEL. 9090 BRYANT
NEW YORK, January 10, 1916.

J. P. Morgan, Esq.,
23 Wall Street,
New York City, New York.

Dear Mr. Morgan:

The war has given special interest to the manufacture of nitric acid from the air by electrical processes. Heretofore, but one hundred thousand tons were consumed annually in all the world's industries. Many times that amount is now required for explosives only. The acid is already very scarce, the market price has trebled and is rising rapidly.

Everybody naturally thinks of water-power, but the fact is that even under the present conditions first class hydro-electric power is too dear. When a plant at Niagara was started years ago with power at $15.00, I advised my friends that the undertaking would surely fail and it did. But there is plenty of waste power in the United States which can be very profitably employed in the production of nitrates. I have pointed out the possibilities of such manufacture in connection with the steel industry in the article inclosed, a typewritten copy of which was forwarded to your Long Island home at an earlier date. Years of careful investigation and a number of inventions which I made in their course have preceded these statements. There is a marvelous opportunity. The Steel Corporation could secure immense revenues by applying my improvements. They are making millions out of my inventions already. Is there no way in which I might be a little assisted in making a practical demonstration?

Yours most faithfully,

N. Tesla

Inclosure.

In this typed letter to J.P. Morgan Jr., from January 10, 1916, Tesla's letterhead still bears an image of his ill-fated Wardenclyffe Tower, but the surrounding technologies emphasize his range as a practical inventor.

changing visionary—in early 1922 Tesla once again appealed to Jack Morgan for $35,000 to launch a new business based on the speedometer technology (an "air-friction speed indicator"), which was producing up to 3 million of the units annually. Tesla pledged one-third of the Waltham contracts royalty to Morgan as an enticement. As with his last contacts with Morgan, however, the relationship proved sterile.

Meanwhile, Tesla's reputation continued to garner worldwide attention and draw the inventor into precarious political situations. The entire core of Western civilization was being threatened by the rise of communism and fascism in Europe. Although millions of Americans didn't want to have anything to do with another massive European conflict, the government didn't want to be caught flat-footed if the growing storm crossed the Atlantic (or Pacific), and the New York press courted Tesla for his weapons systems idea.

Vladimir Lenin, having created the Soviet Union with the Bolsheviks, wanted Tesla to build an AC system in Russia—even as Russians were starving. Federal agents were watching Tesla and the shadowy groups that tried to lure him into the communist sphere. Yet the gaunt inventor was more interested in late-night walks and the welfare of New York's pigeons.

Edison, in contrast, continued to bask in his fame as others continued to maintain his celebrity, even though he was tinkering with failed ideas like converting goldenrod sap into rubber while he communed with his admiring friends Henry Ford and Harvey Firestone. In his final years he attempted to rehabilitate the image of Thomas Paine, the pamphleteer whose reputation as a fomenter of the American Revolution had fallen into decline. In his 1925 essay, Edison lauded the author of *Common Sense*, stating that Paine, who designed an iron bridge and a hollow candle, should also be remembered as an inventor: "He was interested in a diversity of things; but his special creed, his first thought, was liberty." History would vigorously herald Edison's inventions long after they had been replaced by better, more efficient technologies. In fact, major utilities that had "Edison" in their title had employed Tesla's technologies and not the DC systems that Edison advocated in vain.

While the 1920s kept progress surging in the form of a consumer culture built around the modern electrical age, the world saw unprecedented growth in every industry from textiles to consumer products like flatirons. Everyone from housewives to machine tool operators could be 100 times more productive because of power that was being universally supplied by Insull's conglomerate and those controlled by GE and Morgan affiliates and other powerful regional

The World Magazine, January 30, 1916

TESLA DESCRIBES WIRELESS WARFARE OF THE FUTURE

If Tesla's secret invention fulfils expectations it will be possible to blow up battleships at a distance of 1,000 miles from the aerial station.

For Gunpowder Will Be Substituted Electrical Waves Transmitted Over Great Distances and Focussed on Given Targets—Aerial Projectiles Will Be Exploded From Wireless Stations—Whole Populations May Be Wiped Out By Pressure of a Key.

By Edward H. Smith.

IN the imagination of every electrical expert in the world—from the famous Marconi, Edison and Tesla to the prospective aspirants to glory—there is to-day one vision. This is that of a little machine with a key by means of which a wave of electric energy will be flashed through the air to explode the enemy's bombs, torpedoes, cartridges and magazines.

The man who will perfect a device whereby an electric wave can be aimed and steered through the air in one direction only, all its force concentrated on a given target, will go down in history as the greatest inventor of all time, for his machine will make gunpowder and dynamite obsolete and will send rifles and cannon to the junk-heap.

In an interview published in a Paris newspaper recently Marconi speculated upon the possibilities of such an invention, saying it would mean the abolition of firearms and a reversion to hand-to-hand fighting.

A Dutch engineer named Lanzius, now in New York, claims to have invented such an apparatus. He is by no means the first to make such a claim. An Italian engineer won for himself much notoriety two years ago with his demonstrations of a machine that seemed to do just that, until he was exposed as a fraud. A young New Yorker, who already has several valuable and authentic inventions to his credit, claims to have perfected a way of emitting a wireless electric current that will instantly melt all metals within a certain radius. He, however, has been unable to demonstrate it, except on a small scale, owing to the difficulty of banishing every bit of metal from the persons of the experimenters. These would have to wear buttons and studs of bone, suspenders with fabric buckles, shoes with wooden nails, and would have to lay aside all money, pocket knives, steel or gold-mounted eyeglasses, watches, chains and scarfpins, and their teeth would have to be innocent of gold, silver or amalgam fillings!

TO DAY the Russians report that "the Germans have melted down their barbed entanglements from distant trenches." Yesterday a Californian pretended to defend the country from invasion by wireless energy.

The electrical wizards whose names are household words admit that they are busy along just these lines, but they refuse to commit themselves as to the precise object of their research. It is a race between experts and the side whose experts first discover the secret of directing powerful wireless waves will win the war, for it will mean annihilation for its enemies on land and sea.

The idea is real and fascinating; it has the romance of the startling possibility. Who? What? How? I have just talked with Nikola Tesla, that genius product of Serbian soil and Austrian training, who has just been awarded a [illegible] in this year's Nobel prize for physics, surely one of the first inventive imaginations of the world, a unit of the miraculous. And he has told me that:

1. Probably no revolutionary invention will decide this war, unless one counts the provision of internal needs in Germany by scientific substitution.

2. This is the last war in which gunpowder will decide the issue.

3. The whole aspect and conduct of war will change. Electricity will be the force of organized murder to-morrow.

4. It is not unlikely that that ignis fatuus, an invention so terrible as to make war impossible, will be found. Has been found, perhaps!

We sat in the inventor's offices, near Fifth Avenue, looking at a picture of his new wireless telephone plant, designed to develop unlimited, undreamed of electrical energy for wireless transmission, a Merlin's tower, overshadowing the imagination. From the desk Lord Kelvin watched us from his portrait, and there gleamed elusively a thing the size of a watch and the power of ten horses.

"The next war will be fought with electricity," he said. "Cannon will be impotent compared with what is to come."

In other words, the Promethean struggle now taxing the vitality of a hemisphere will mark the end of the potency of gunpowder. The invention of Schwartz and Bacon, which first unthroned the armored knight at Crecy in 1346, if there be not some fiction in history, was to give way after 500 years of dominance in the affairs of men and their nations.

The career of this old monarch of mortal destinies has waxed through the centuries. His kingdom has become ever wider and more terrible. The conquering Turks used a cannon or two, hurling 600 pound stones, in their sieges of Europe's capitals. At Marignana, in 1515, the French under Francis I. fought the first real artillery battle of history, defeating the Swiss and Milanese with his 350 primitive cannon. Since then battles have come to be more and more, the applications of gunpowder to great arms. Now whole nations devote their complete industrial resources to feeding these cannon, to supplying their deadly loads. "The limit has been reached. War can be no worse," say the hopeful.

"The limit has no more been reached in the great cannon than in the two-handed sword," says Mr. Tesla. "The world will merely change arms. But the change will not come in time to decide this war, because it takes time to develop such revolutionary things."

I SHOWED the inventor a clipping reporting that the Germans had cut down Russian entanglements with heat, supposed to have been applied from their trenches by the use of reflecting mirrors. It gave the impression that troops themselves might be thus incinerated from afar.

"In my opinion," said Mr. Tesla, "the melting of barbed wire from a dis-

Inventors are working to perfect a device which, by the emission of wireless currents, would explode the cartridges in soldiers' belts.

The wireless transmitting tower which Nikola Tesla built at Wardenclyffe, N. Y., and in which many of his wireless experiments have been conducted. The tower is 185 feet high.

Once an electrical wave could be guided through the air with all its energy concentrated on a given point munition plants could be wiped out by the score.

The operator's tower as it might be at Tesla's wireless station. Below are the aerial projectile tubes trained on battleships hundreds of miles distant. The projectile—flying many times faster than an aeroplane—is exploded by electrical waves as it reaches its target.

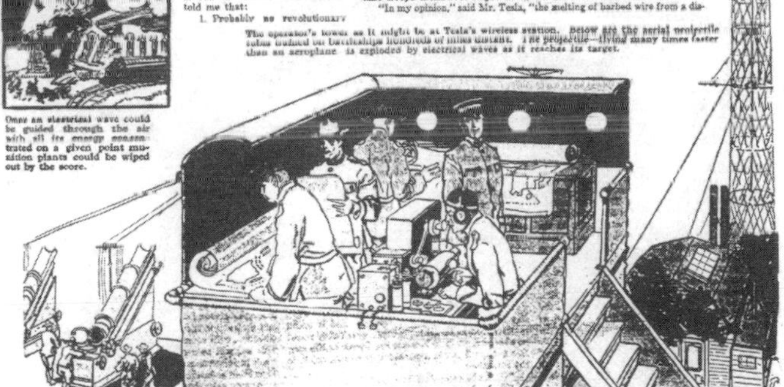

An article in *The World Magazine* from January 1916 features Tesla's early ideas for applying his wireless technology to national defense—ideas he continued to push as Nazism became a serious threat to Western civilization and that foretold the advent of modern drone warfare.

This illustration by Frank R. Paul from a 1922 edition of *Science and Invention* reveals the state-of-the-art electrical infrastructure of the (science-fictional) "Apartment De Luxe."

combines. It was a new, go-go age as people went to work in offices instead of factories and farms, and rode in elevators instead of horse-drawn carriages.

Yet there was a dark undercurrent to that time. Hate groups like the Ku Klux Klan rose even in northern states like Indiana, and industrial icons like Henry Ford spouted anti-Semitic diatribes. The stock market created another mania. Organized crime increased during the "dry" years of Prohibition. Whatever new wealth was being generated in the world, it wasn't being distributed equally—particularly in Europe, which was cowering economically from the aftereffects of the Great War. The dragon of disenfranchisement was coming out of its cave, and Tesla's "World System" was increasingly seen as an asset to protect democracy. Although by that time he was well past his halcyon days as a practical inventor, Tesla continued to promote his system. Emphasizing its potential for facilitating world peace—in addition to protection—gave it even more currency.

Tesla's Last Inventions

Well into the 1930s, Tesla was still trying to build a working relationship with Westinghouse, even though it had soured in 1930, when the inventor accused the company of infringing on his patents. Still fighting patent battles in the courts, Tesla was receiving some royalties from his earlier inventions, including the speedometer, in the 1920s. Seeing the popularity of radio go viral with mass electrification of homes, he must have been supremely aggrieved that he wasn't making a dime off his fundamental technology.

Tesla's last patent, filed in 1928 when he was seventy-two, had nothing to do with electrical engineering per se. Hoping to improve on the design of the helicopter, which had been invented in 1921, Tesla created drawings of what appears to be a vertical takeoff and landing (VTOL) vehicle, perhaps the first of its kind since the days of Leonardo. What's interesting about Tesla's VTOL isn't so much its design, with his bladeless turbines, but that it doesn't have an onboard power source. It was designed to be powered by a network of Tesla's magnifying transmitters, the first of which he still hoped to build at Wardenclyffe.

"Aerial machines would be propelled around the world without a stop," he had originally predicted in 1900, when he was just breaking ground at Wardenclyffe. The inventor, going even further into the future, imagined a portable version of his VTOL that would weigh under 250 pounds and could be "run through the streets and put in a garage."

The giddiness over universal air travel embraced the nation shortly after Tesla filed his last patent. The year before, Charles Lindbergh had flown nonstop from the East Coast to Paris, becoming an international hero. The modern age was not only in love with electricity; it was lapping up the notion of unlimited mobility. Why sit on sluggish steamships or trains for days when you could fly somewhere in a matter of hours? In keeping with Einstein's enhanced view of the infinite—limited only by gravity and the speed of light—why couldn't people travel anywhere anytime they wanted, perhaps even entering outer space? That was the promise of modernity in the late 1920s.

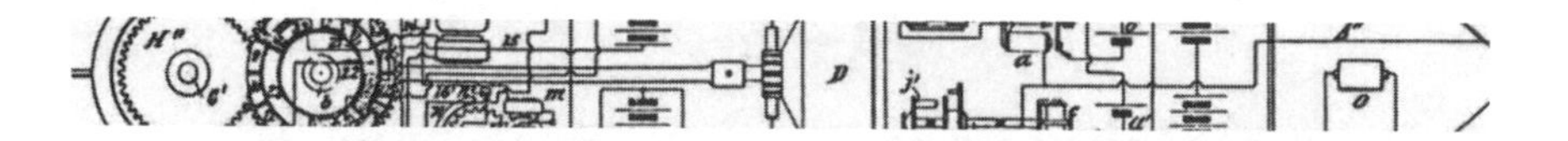

TESLA, BOWIE, AND *THE PRESTIGE*

As I was divining the significance of Tesla's Colorado Springs discoveries, I came across Christopher Nolan's magnificent 2006 film *The Prestige,* based on the novel by English author Christopher Priest. The plot centers on two vindictive magicians practicing their art in the late nineteenth century.

After the wife of one of the magicians—Robert Angier, played by Hugh Jackman—is killed in a sabotaged trick, the two showmen spend the rest of their lives trying to foil each other's illusions while making their own even more spectacular. Angier, who blames the other illusionist (Alfred Borden, played by Christian Bale) for his wife's death, is bent on destroying his rival.

Both magicians want to showcase the ultimate deception: to disappear from stage and suddenly reappear somewhere else. While I won't spoil the plot—you'll want to see the film several times to enjoy its delightful mysteries—the story takes a thrilling turn when Angier travels to Colorado Springs, where he meets Nikola Tesla experimenting in the mountains. Tesla is played by the immortal chameleon David Bowie (who died just as I was writing this chapter).

Although Bowie's resemblance to Tesla is less than ideal, there might have been no better match for the inventor than the iconic rock star. Both men had an aura of mystery in their various personas, were extremely popular in their heyday, and inspired millions of others.

Bowie, who was feted in a Chicago Museum of Contemporary Art exhibit in 2014, was the ever-changing cultural chimera. The rock star sang about traveling to the heavens, and then morphed into a glam-rock hero, and then transformed into an R&B singer, followed by ventures into electronica, pop, jazz, and cabaret. His numerous incarnations—even as he lay dying—left his contemporaries breathless. There was little question that Bowie was an innovator of the first rank, even infusing the world of fashion, theater, and film with his radical new ideas of self and invented persona.

David Bowie performs as the Thin White Duke—one of many personas over his long career—in 1976.

In real life, Bowie went much further than most rock heroes in reinventing himself and the world around him. Sensing that the recording world was moving away from vinyl and CDs, he offered the first downloadable single in 1996. It would have taken some eleven minutes to fully download back then, but Bowie did not let that deter him. The cultural chimera also pioneered derivative securities based on future royalties from his music, later called "Bowie bonds." He set up his own Internet service provider in 1998, even giving away twenty megabytes of storage so that his fans could set up their own homepages.

As the currents of fashion, music, and theater changed, so did Bowie, transforming himself into a music producer and mentor, always advocating for those on the fringes of "normal." My brother Tom, an artist and substance abuse therapist, called Bowie's music "the soundtrack of my life."

When Bowie comes on screen as Tesla, promising the magician he will create a new machine for him in exchange for a princely sum of money, you can feel the charisma of the artist. Although Tesla in real life likely never made such a piece of science fiction, the film depicts the inventor as an illusionist who insists his high-voltage teleportation device "is science, not illusion." The novel on which the movie was based, which described Tesla as "an associate of Thomas Edison" in 1892, did not portray him accurately (Samantha Hunt's *The Invention of Everything Else* contains a more faithful—though fictional—portrayal of the inventor). Factual errors aside, Bowie lent gravity to a character that many were and are still trying to understand. The artist, like the inventor, was a man for all seasons.

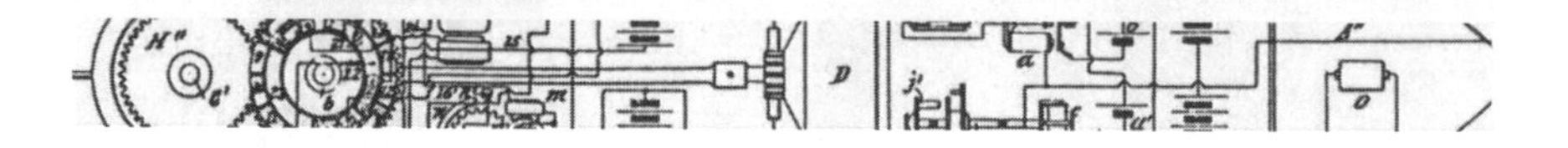

The World Unravels

By the end of 1929, optimism became a rare sentiment as stock, commodity, and property prices crashed and the world entered the Great Depression. An industrialized world once brimming with hope, freedom, convenience, and mobility tried to stave off despair on a massive scale as millions lost their jobs.

Just after the market crashed, Insull was celebrating the opening of the new Chicago opera house, which he had helped finance and build. He owned the penthouse in the building and stayed there after performances. Despite his efforts to keep the stock of his sixty-five utility holding companies afloat, they were all connected like a fragile jigsaw puzzle. Although Insull was on his way to utter financial ruin as his holding companies collapsed and their stock became worthless, he continued to support Tesla financially. According to letters between the two men from the end of 1929 to the fall of 1930, Insull sent a total of $4,000 in checks to Tesla. In inflation-adjusted 2016 dollars, that was the equivalent of nearly $57,000—a stunning amount considering that about one quarter of the American workforce became jobless during the height of the Depression. The average U.S. family income in 1930 was just $1,500.

This 1931 photograph shows unemployed men queued outside a soup kitchen opened in Chicago by Al Capone. The storefront sign reads "Free Soup, Coffee & Doughnuts for the Unemployed."

Tesla, although unable to cement a profitable business relationship with Insull, remained an emotional supporter as the utilities baron became one of the most reviled men in America. Before the crash, Insull was a business celebrity whose Midas touch turned everything from power companies to railroad lines to gold. People would approach him on the street for stock tips. Imagine a combination of Bill Gates and Warren Buffett. That was Insull's image before 1930.

But since thousands of people now held worthless stock in Insull-branded companies, Insull became much more infamous than Al Capone. At least Capone donated money to churches; Insull was believed in the popular imagination to have purloined the money from stock sales and squirreled it

Fallen electricity titan Samuel Insull reads the *Daily News* story of his return from Europe to face trial in Chicago, 1934.

away as the nation started to starve. President Roosevelt singled him out as a prime malefactor in causing the Depression in a 1932 campaign speech that laid out the New Deal.

Insull's innovations in selling stock directly to the public contributed to his undoing. It was Insull who not only pioneered the idea of "publicly owned" utilities—he peddled the stock of his utility companies door to door in the Midwest—but championed the state commissions that "regulated" utility rates, a unique quasi-governmental model that exists today. Yet Insull was severely overleveraged, and his incestuous pyramid of holding companies held stock in each other. The stock values of these mostly paper companies evaporated after the 1929 sell off.

Fearing that he was going to be roasted at the stake in Chicago, Insull fled to Europe, eventually landing in Athens. FDR pressed hard on extraditing Insull, who had since been indicted for fraud, eventually nabbing him in Turkey and bringing him back to Chicago for trial.

Insull was initially tried in the same courtroom and by the same judge who had presided over Capone's tax evasion trial, although the public animus against Insull was exponentially greater. But after three trials, Insull, his son, and several associates were acquitted of wrongdoing. Evidence showed that Insull had gone broke trying to prop up the stock of his companies. Financially, he went down with his ship (unlike most CEOs of today, who walk away from criminal charges with multimillion-dollar platinum parachutes).

In June 1932, just around the time Insull was the center of the biggest business bankruptcy in U.S. history (the Lehman Brothers of its time), Tesla sent him this telegram:

> “The world is enjoying the inestimable benefits of your genius and enterprise. Enough glory for any man. I wish you long life and happiness from all my heart.”

One of the last letters Tesla sent to Insull in 1935 relished Insull's acquittal in the three fraud trials. Tesla wrote that he "rejoiced in the decision" and that Insull was "a great man who has rendered to the world services of inestimable value."

More than thirty years after he entranced audiences with his Eggs of Columbus, remote-control boat, and magnifying transmitter, Tesla was featured on the cover of *Time* magazine.

Was Tesla's praise overdone? Both men *believed* they would rebound after their respective catastrophes. While the public had entirely dismissed their accomplishments and struggled with titanic woes in the late 1930s, both men retained faith in themselves—and each other.

Tesla, now virtually alone in his quest to revive his fortunes, railed against Marconi, who discouraged the use of Tesla's motors in powering warships. Belittling Marconi's first transatlantic signal as a "paltry engineering achievement," Tesla once again promoted his "World System" in a letter to the editor in the *New York World*.

A young science editor named John O'Neill at the *Brooklyn Eagle* took note of Tesla's tirade and took it upon himself to be Tesla's advocate. Some two years later, O'Neill would fete Tesla and his "epoch-making" inventions, while quoting the inventor on his claims that some waves could exceed the speed of light and that he could "harness cosmic rays to power a motive device." O'Neill would also write the first biography of Tesla.

Such attention to Tesla's most challenging ideas certainly raised the inventor's public image, although not always in a positive fashion. Now other, less sympathetic journalists were attracted to the inventor for the mad scientist persona they wanted to sell. Tesla's pronouncements during the Depression became part freak show, part prophecy.

Tesla's renewed fame ascended to such a degree that when he turned seventy-five in 1931, he was featured on the cover of *Time* magazine in recognition of his myriad contributions to uplifting humanity from darkness at a time of widespread loss of hope. Birthday greetings and congratulations

GREAT SCIENTIFIC DISCOVERY *Impends*

—*Nikola Tesla*

"I Positively Expect That Within the Next Decade New Sources of Energy Will Be Opened Up Which Will Put at the Disposal of Mankind Everywhere Power in Unlimited Amounts—I Have Made a Discovery Which I Expect to Announce Soon"

By Harry Goldberg

NIKOLA TESLA, the noted inventor, is preparing to make public within the near future the announcement of a discovery which will make available to the world new sources of energy.

Men Are Like Bees

LEISURE—"Too much leisure, and civilization will go to pot. Man was born to work and suffer and struggle; if he doesn't, he will go under."

MATERIALISM—"Technical advances are inevitably driving us toward the grossest kind of materialism. Before long the social system of bee life will become universal."

LIFE OF PLANETS—"There must be life on other planets. The sun shines. The stars give out heat. Water collects on the surface. Chemical changes occur that we do not yet understand."

—NIKOLA TESLA.

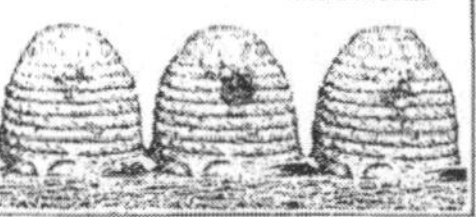

At the ripe age of seventy-five, Tesla still had his visionary focus, making a number of predictions about the future of humanity in this 1932 *Galveston Daily News* article. Among them: "Technical advances are driving us toward the grossest kind of materialism. Before long the social system of bee life will become universal."

poured in from around the world—even from Albert Einstein, who chose not to debate Tesla on the limitations of the speed of light.

When Tesla turned seventy-six, he said he believed in life on other planets, adopting a more expanded view of the cosmos. The language he evoked in his writing of that period became decidedly spiritual, although he never embraced a particular religion. Looking beyond his role, he asserted that mankind would soon have the capability to "alter the size of this planet, control its seasons, guide it along any path [it] might choose."

By 1935, when Tesla had written his request to the then-ruined Insull on his funding needs for telegeodynamics, the world was hurtling into a war between fascist-totalitarianism and social democracy. Tesla was mentioned as working on "his greatest achievement" that year, according to the *New York Times*. Although the description is vague, the inventor bragged of "an apparatus by which mechanical energy can be transmitted to any part of the terrestrial globe." Not only could this device act as a geographic location system, but it could locate ore deposits and create "controlled earthquakes."

As I noted earlier, Tesla thought his device could harmonize with the earth's resonant frequency to send high-speed waves through it. The inventor even boasted that with such technology he could bring down the Empire State Building with "five pounds of air pressure."

Despite his financial state, it had been a productive decade for Tesla and his dream devices, which included a description of a geothermal energy system and his "machine to end war." With another "radiant energy" apparatus, which spurred the "free energy" movement that's still alive today, he hoped to tap the abundant energy flowing from the sun and electromagnetic radiation from other stars.

Of course, the end of the 1930s was a time in which everyone was on edge because of Hitler's evil juggernaut and the sympathetic dictators Mussolini and Franco. What did the dictator plan to do with the greatest war machine ever built? Tesla countered that he was sure he could prevent any attack with his weaponized "World System" that could repel "10,000 planes or an army of a million"—from 200 miles away. Tesla offered such a system to the British government in 1936. His Majesty's War Office chafed at the cost of tens of millions of pounds and moved on.

Sensing that his system could be used to defend free society, Tesla openly pitched it as a base for his electrical weapon, which the papers had dubbed the "death ray." For many who had been following Tesla's career—particularly in the federal government—it would be the last thing they remembered about the inventor.

TeslAction 8 Reposition Yourself

Just as Tesla repositioned himself as an all-purpose inventor in the early 1920s, it's often necessary for people to reinvent their brands when confronted with setbacks such as economic recessions. Writers, musicians, and reformers often go through this rite of passage. Having written his great English novels in the late 1800s, Thomas Hardy focused primarily on poetry in the early twentieth century. William Butler Yeats moved from poetry to national politics in the 1920s. Activist Jane Addams shifted from serving the poor to lobbying for peace, becoming the first American woman to win a Nobel Peace Prize in 1931. Frustration is inevitable when you're trying a new avenue, though. Not everyone can design a new cellphone. Give yourself some room to fail. The next big thing might not work out, but there's always tomorrow.

When he started to feel his relevance fading, what did Tesla do? He staged media events in his sixth decade and beyond to herald his new ideas on his birthdays, inviting everyone in the New York press. He came up with new ideas. Although he didn't have a lab or money, he still had intellectual capital aplenty. As world peace became more endangered, Tesla was the one who attempted to marry technology with a common, peaceful purpose. He transformed himself from genius inventor to thought leader. The man who had "seen" the rotating magnetic field was talking about spaceships. Would he have been out of step in our own time when visionaries like Jeff Bezos, Elon Musk, and Richard Branson are reaching for the heavens with their billions and branded rocket systems?

After being abandoned by New York's finance community, Tesla transformed himself into an éminence grise of the intellectual frontier. His and countless other examples show that it's possible to rebrand yourself well into your life and career. After my family's most challenging years (2008–2010), I repositioned myself as an advocate for investor protection, starting a blog ("Bamboozlement") that eventually pulled in up to one-half million views per month. Supplementing my income with a full schedule of speaking, book writing, and freelance journalism, I realized that creating a platform for my advocacy work was extremely important.

While my blog income at the time was minuscule compared to my former precrash income—it would take a month's worth of blog income to equal what I made in one hour writing global financial columns once a week—I had to keep an open mind, accepting the new media environment and adapting to it. It was painful at first, but I soon realized that I just needed to present myself in a different way: I was an entrepreneur of ideas. What could I bring to the world that was unique or useful? What everyday concerns that few people wanted to talk or read about could I address?

While Tesla never lost his vision of the "World System," he kept asking those questions of pragmatic concern. Look at what's needed in your community and the world at large. Is there something you want to pursue that could help your community? Can you develop a new application that makes people's lives easier or eases their pain?

For those of you who've suffered from job loss, take heart and refocus on smaller projects that may bring some money in the door while you try to find a new approach. As an exercise, write down the things that make you feel most alive. What does it take to get your spirit recharged? Some of them may be translatable into a new vocation or simply a different way of looking at your human capital—the skills that you possess and can contribute to the world in a unique way. Ask yourself:

- **What do I do especially well that I'm passionate about? Can I find a way to make money at it?**
- **What niche can I fill? What needs to be done, made, or offered that I can bring my skill set to?**
- **What is my course? Do I need to take new bearings?**
- **How can I better connect with the world? What do I need to know?**

At the very least, stop thinking about finding a job. Like Tesla and Einstein, you can create a new opportunity for yourself. With the omnipresence of Internet services and social media, you can wirelessly broadcast your service to the entire world.

WRNY
AMAZING STORI
HUGO GE
EDIT
Stories by
H.G. WELLS
A. HYATT VERR
JULIAN HUX
XPERIMENTER PUBLISHING COMPANY, NEW YORK, PUBLISHERS

IX

A DETECTIVE STORY

SEARCHING FOR THE ELUSIVE DEATH RAY

Nikola Tesla shared with humanity the most important message ever communicated. At the same time, he was an uber-communicator; his way was a complete process where thought grows from the visualization of an idea to discovery, where a message develops through improvement of existing and invention of the yet unknown. Tesla's message is yet to be received by the momentum of tomorrow.

—Nevena Vukasinovic, ENGSO Youth Secretary General, Serbia

It was as if life itself had cruelly scooped the vitality out of Tesla's once-handsome face, his weary eyes sunken into the crevasses of his broad forehead. When his young relative William Terbo saw him, though, he was still effusing happiness at seeing kin, hugging the boy and kissing him three times in the Serbian way. The old man, now willow-branch thin, tousled William's hair with his hand.

What was this? The man who famously refused handshakes and bodily contact for fear of germs was physically affectionate.

"Tesla had treated me differently than I expected," Terbo told me. "I think he saw in me a reflection of him at my age. We both had older brothers who

The August 1927 cover of *Amazing Stories* depicts H.G. Wells's 1897 sci-fi masterpiece *The War of the Worlds,* in which Martian invaders incinerate everything in their path with deadly heat-rays.

An elderly Tesla is photographed in his hotel residence, c. 1940.

had died tragically at a young age. It was a catastrophe for us, losing my brother Jackie. He must've known how I felt."

William, all of ten years old, was spending a week at the Jersey shore and then visiting New York City with his mother, Alice. It was toward the end of the Great Depression, and Tesla was in his last decade, his faded celebrity on a slight upswing after he had been interviewed years earlier in several publications and appeared on the cover of *Time* in 1931. He was now living in modest rooms in the Hotel New Yorker with his dreams, inventions, papers, and pigeons to keep him company.

Terbo's father, Nicholas (Trbojevich), was a successful engineer and inventor in his own right (most importantly of an automotive hypoid spiral bevel gear design still in use today). He had been forwarding money to his uncle Tesla, including at least $2,500 and possibly other payments in the form of "gifts, loans, or investments," but Terbo thinks he stopped his financial aid in 1937.

Before a visit to the sparkling entertainment palace at Radio City Music Hall, mother and son were briefly meeting with the inventor. Terbo recalls that his mother had some family business to discuss with Tesla, although he wasn't sure what it was. The meeting lasted less than an hour, and they were on their way. William wasn't quite sure what had transpired.

Like his father and granduncle, William Terbo had a fertile career as an engineer, originally designing rocket equipment in the U.S. space program, then moving into telecommunications. Where Tesla had dreamed of tapping the vast energy of the cosmos, Terbo would help put astronauts into space, designing explosive bolts that would separate rocket stages. "I was predestined to become an engineer," Terbo said. "My father said you couldn't go wrong as an engineer. You'd make three times as much as an electrician."

What still resonates with Terbo—as if one of Tesla's massive magnifying transmitters were still oscillating Tesla's life energy—is the inventor's compassion and empathy. They were two souls adrift in the world without their beloved big brothers. "When he saw my mother and me together for the first time since Jackie's passing, I believe he flashed an image of his mother and himself," Terbo recalled. "He patted me on the head and messed with my combed hair."

As the last living person to have met Tesla, Terbo is the guardian and official caretaker of the Tesla legacy. You will often find him, as I did, at the key conferences honoring Tesla and exploring his ideas. I first met Terbo at the Hotel New Yorker at a Tesla conference, where he spoke briefly with great gravity and dignity. Terbo has spent more than forty years preserving and enhancing the public's knowledge of his granduncle as the chair of the Tesla Memorial Society Inc. He's been present to dedicate many statues and memorials recognizing the inventor's achievements.

Terbo, who has an archive of papers and articles on Tesla to rival most libraries, dismisses the idea that Tesla was mentally unbalanced, not interested in women, or even aloof. In addition to Tesla's familial warmth, Terbo champions Tesla's creative abilities, encouraging more people to study his work and be inspired by him.

"Tesla could imagine everything he needed," Terbo added. "He could keep things in his mind with remarkable clarity."

The End Comes

The inventor, even in his frail state, had been recognized by a host of countries in the years leading up to his death. Two years before Hitler ravaged Poland in 1939, Tesla was honored on his eighty-first birthday in New York City, where the minister of Czechoslovakia presented him with the Order of the White Lion and the ambassador of Yugoslavia awarded the Grand Cordon of the White Eagle, along with a $600-a-month pension, on behalf of King Peter II.

Later that year, after leaving the Hotel New Yorker on one of his nightly jaunts to feed his pigeons, Tesla was hit by a cab. Even though he was thrown as far as forty feet by some accounts, he refused to go to the hospital and instead hobbled back to his hotel. Tesla probably suffered broken ribs and was never in decent health after that incident, mostly confining himself to his hotel room from then on.

Dominated by a parrot-like beak, the inventor's head seemed much too big for his body when young King Peter II came to visit him on July 8, 1942. He was as thin as one of his canes, yet he was still seen as a robust hero in the Balkans. Before Hitler invaded his homeland in 1941, it was Tesla's name the Serbian people had whispered when all hope seemed lost in the face of a fascist invasion. After all, Tesla said he would help shield *all* peaceful nations willing to adopt his "World System." It was Tesla, the expatriate folk hero, who gave the world ubiquitous power. It was Tesla who knew the secrets of this unseen, god-like force that could repel waves of invaders—even the dreaded Luftwaffe, which had subdued Poland by a blitzkrieg in a matter of days.

In a telegram to his nephew Sava Kosanovich on March 1, 1941, Tesla wrote of a neutron-based weapon that could "destroy the largest ships afloat.

With the assistance of his nephew, Sava Kosanovich (third from left), Tesla meets the youthful King Peter II (second from right) the year before his death in 1942.

There is unlimited distance of travel. The same is for airplanes." To deploy the neutron weapon, Tesla proposed at least six stations in the Balkans: one for Serbia, three for Croatia, and two for Slovenia. Although he didn't provide any details on how such a system would work, he suggested that it could be powered by 200 kilowatts, "which can defend our dear homeland against any type of attack." The weapon system was unbelievably ambitious, considering there wasn't a particle accelerator in existence that could produce enough energy to create that kind of weapon. Tesla further suggested in another telegram to Kosanovich three days later that "electrical energy will deliver particles through space with the speed of 118,837,370,000 centimeters per second. This is 394,579 times the speed of light." (At this point, Tesla said this weapon could be powered with 20 million volts.)

With Hitler, there was no negotiation. You either capitulated or were staring at the Wehrmacht or other Axis fascists goose-stepping down your streets with the people who hated you the most (in Serbia's case, Croatian fascists and Bulgarians). Tesla knew his countrymen were imperiled and once again promised that his technology could help them, although by the time King Peter II visited him in New York, it was much too late for Europe. The monarch of Yugoslavia had originally come to New York to plead with Eleanor Roosevelt and others to save his country, but the Allied forces had thrown their support behind the Yugoslav communist rebel Josip Broz Tito. Visiting Tesla briefly in his hotel, the king was appalled at the inventor's state of decay. They reportedly wept together, as there would be no saving their homeland.

Was Tesla mentally competent in the last year of his life? It was clear that the inventor was nearly destitute and was suffering from a number of maladies as his body broke down. His isolation also didn't help. Yet what remained of the free Balkan countries wanted Tesla to be an avenging angel, a lone ranger who would ride in with his technology to save the world from the evils of fascism. While he was writing articles about world peace and feeding his pigeons, Tesla's heart began to fail and he suffered from fainting spells.

A drawing of the first nuclear reactor in the West Stands section of Stagg Field at the University of Chicago. The reactor consisted of uranium and uranium oxide lumps spaced in a cubic lattice imbedded in graphite.

As the conflagration of World War II grew, a weapon of vast destructive power was enabled when Enrico Fermi and his team of physicists created the first sustained nuclear reaction in a "pile" of uranium oxide and other heavy metals under the football field stands at the University of Chicago on December 2, 1942, a month before Tesla's death and literally walking distance from where Tesla had gloriously showcased his AC systems some fifty years earlier. An abstract sculpture of a mushroom cloud marks the spot, which sits as a bereft, ominous monument next to the ultramodern cocoon of the Mansueto Library.

Tesla's weapons would have to wait while the inventor slouched closer to eternity.

• • •

It was a maid who discovered Tesla among the pigeon dung on January 8, 1943. She immediately alerted management, which later summoned a locksmith to open Tesla's safe in the presence of government agents from the Office of Alien Property (OAP), Sava Kosanovich, Tesla's friend Kenneth Swezey, and George Clark, the director of an RCA museum.

Abraham Spanel, president of the International Latex Corporation, who was interested in Tesla's "death ray" pronouncements, voiced his concern to the FBI that the inventor's weapon might fall into the wrong hands if his plans made it back to Yugoslavian fascists through Kosanovich. The *New York Times* ran an obit with the subhead "Claimed a 'Death Beam'/He Insisted the Invention Could Annihilate an Army of 1,000,000 at Once."

Just after Tesla died, John Trump, MIT's director of high-voltage research, reportedly combed through Tesla's papers for any hint of a usable weapon. His conclusion was straightforward: "Nothing of value for the war effort and nothing which would be helpful to the enemy if it fell into enemy hands."

Despite this expert opinion, FBI director J. Edgar Hoover wouldn't let go of the idea that Tesla's weapon ideas were, in Trump's words, "of a speculative, philosophical and somewhat promotional nature." Well into the 1950s, Hoover was probing whether there were any communist connections to Tesla's work and associations.

COPY LOUIS ADAMIC . MILFORD . NEW JERSEY

January 4, 1943

Dear Mr. Hoover:

Nikola Tesla, as you know, is a Serbian immigrant who came to America from Croatia some 60 years ago and became one of the world's greatest inventors. He became also an American. In the early 1920s Lenin urged him to move to the Soviet Union, promising him every scientific facility, and personal security for life, but Tesla declined -- he was an American and had got used to living in the United States, whose civilization he had helped to create.

His contribution to the sum-total of American civilization is almost beyond calculation. Hundreds of billions of dollars of American wealth are ascribable to his inventions. They are at the very center of our current war effort. No man living has added more substantially to the potentialities of human life than Tesla.

Yet today, when he is past 90, he is worse than penniless. He is extremely frail, weighing less than 90 pounds. His health is poor, and he has grown somewhat bitter against the U.S.A. No doubt his current poverty is his own fault. However, I think that ordinary standards do not apply to Tesla. He was always the pure scientist, never interested in money, always impractical about material existence.

But the fact is that now he is up against it. He receives a small "pension" from the Yugoslav government-in-exile. I know that Tesla suffers greatly at having to accept this pension from the government of his native country, to which he had never contributed anything directly. He suffers especially because the money comes to him through the Yugoslav Ambassador in Washington, whom he dislikes personally. Tesla suffers, too, in fact to the point of bitterness, because he feels -- with some justice -- that everyone in America, including the beneficiaries of fortunes created by his inventions, has forgotten him. No one writes to him; no one comes to see him.

He lives in a meager room in the New Yorker Hotel, in New York. He owes about a year's rent -- the Yugoslav pension is not enough to keep him in scientific apparatus, etc., for he continues to work on his projects.

This letter is not an appeal for your personal financial help. Some way will be found of looking out for him -- he will probably not outlive 1943. But he needs someone to take care of him personally without seeming to; someone who could also follow his current notes and experiments and preserve what may be of value in them. Perhaps one of the large electrical corporations which have benefitted so greatly through his inventions would be glad to pension him for the short balance of his life. And I am wondering if you know someone who might be approached.

A pension coming from such a source would relieve Tesla of the necessity of accepting more money from the Yugoslav government. It would do much to remove his bitter feeling of neglect. And it would be a fitting, though small, recognition of the debt America owes this man who has done so much for his country.

If you would like more details, I can come to see you in New York at any time.

Sincerely, (s) Louis Adamic

Just a few days before Tesla's death, Slovene-American author Louis Adamic wrote a heartfelt letter (shown) to former U.S. president Herbert Hoover, requesting help in securing a pension for the frail inventor. Emphasizing Tesla's extreme poverty and loneliness, he reminded Hoover that "hundreds of billions of dollars of American wealth are ascribable to his inventions" and that "no man living has added more substantially to the potentialities of human life than Tesla."

As for Tesla's legacy as the inventor of radio, Tesla was vindicated by a Supreme Court decision invalidating fifteen out of sixteen of Marconi's patents on the technology. The ruling came a year after Tesla died.

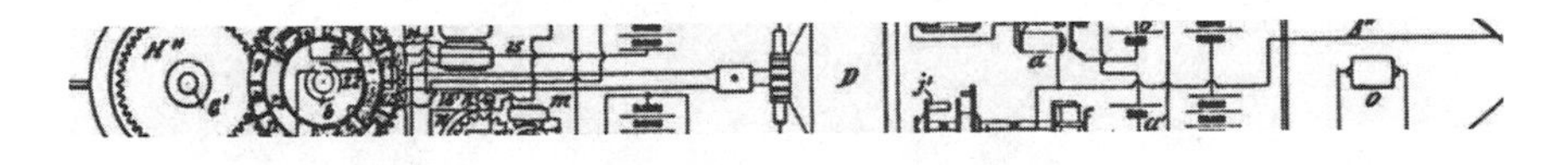

WAR OF THE WORLDS

Having been schooled in the wake of Tesla's visits to Chicago and Milwaukee, a young Orson Welles had an imagination like no other man of theater. Inspired by H.G. Wells, the author of *The War of the Worlds,* he fantasized about Martians getting signals from our planet—and invading.

Welles managed to scare the life out of radio listeners who tuned in to his broadcast on October 30, 1938, with his brilliant depiction of the Martian invasion that mimicked an actual newscast. He also heard in his aural imagination what Tesla-like death rays would sound like. In a stroke of pure theatrical genius, he was able to produce the most terrifying sound that had ever been heard on radio: the sound of an electrical pulse annihilating entire towns.

The son of an inventor with a dozen patents to his name (having invented an automobile jack and bicycle light) and carousing salesman who frequented Chicago's brothels, Welles never saw his father after the age of seven. Following his stunning radio success and creative productions at his New York–based Mercury Theatre, co-founded with John Houseman, Welles was given a carte blanche contract by RKO Pictures with nearly total creative control to make Hollywood movies. (Under the rigidly controlled studio system, this was an exceedingly rare privilege.) Welles's first production, which many consider to be the best film *ever,* was *Citizen Kane.* He created an innovative motion picture in every sense of the word, evoking the power that came into the world during the early twentieth century as he portrayed the too-close-

This photograph shows Orson Welles making his famous *The War of the Worlds* radio broadcast on October 30, 1938.

for-comfort rise and fall of newspaper magnate William Randolph Hearst, although Welles would later claim that he modeled Kane after several industry titans of the time.

The story arc covered more than seventy decades in the life of one man, the fictional Charles Foster Kane, who rose from humble beginnings to head a media empire. As Welles prepared to play the dying Kane, he invoked the image of a man who had fallen from great heights only a few years earlier. He told his makeup artist "to make me look like Sam Insull."

Meanwhile, in the real world, Albert Einstein heard that the Nazis were working on a device that exceeded Tesla's most potent weapon and wrote his famous letter to President Roosevelt, explaining how a nuclear chain reaction "would also lead to the construction of bombs." Einstein urged Roosevelt to form a team to explore the possibility, calling for "quick action on the part of the administration."

The genie of pure mass destruction was about to be let out of the bottle. Who would possess the cork?

Tesla's Contribution to the New Electrical Age

To put Tesla's work into context, it's important to examine the cascade of developments his AC technology triggered that culminated in our modern, hyper-connected consumer culture. One needs to walk back again to the 1920s to see the enormous societal progress that widespread electrical power enabled—advances that slowed considerably in the following decade. After Einstein had unlocked Pandora's box on time, space, and gravity, the world was being wired with pure energy. Cities pulsed with extensive electric trolley systems. Skyscrapers, facilitated by ever-stronger steel frames and elevator systems, got taller. Automobiles and trucks started to clog streets and highways that had been designed for horses and carriages.

Although the interstate highway system didn't fully emerge until the 1950s, transcontinental roads had their roots in the Roaring Twenties. In

1925, the original plan for U.S. Route 30—Lincoln Highway—was put on paper. Eventually, the coast-to-coast route would link Astoria, Oregon, and Atlantic City, New Jersey.

Henry Ford and other automakers, thanks to electrified mass production, were making vehicles more plentiful and cheaper. Electrified interurban trains allowed Americans to live in suburbs and work in central business districts. All of the major eastern and midwestern cities (plus Los Angeles, with its streetcar network) would have these well-connected rail networks by the end of the decade.

But for most Americans, electricity became an everyday reality because their own homes were being wired, with now-familiar sockets installed in more than one room. Gas lamps were typically retrofitted for Edison's electric bulbs. In the 1920s, rapid electrification hastened the wiring of the majority of homes in urban areas, led by Insull's myriad electric operating companies, General Electric, and other regional utilities. And the rate of electrical expansion was exponential: In 1921, one million homes were electrified—a rate that jumped to 2 million by 1924, according to David Nye in *Electrifying America*.

Although you take for granted what you can plug in today, selling power in the 1920s required an onslaught of mass marketing. At first, Insull's utilities sold appliances directly from his own stores, as there were no major retail appliance outlets. Utility companies built model homes that were fully electrified, even in rural areas, to show the virtues of the latest electric toasters, hand mixers, heaters, and cow-milking devices.

Long after the "Battle of Currents" had been decided in Tesla and Westinghouse's favor, GE and all the local utilities promoted AC power for everyone. Housewives were pitched on single, 150-watt ceiling bulbs for six dollars apiece, payable in twelve installments. Insull paraded trucks carrying electric irons down residential streets and *gave them away.* Eager homeowners signed up for electrical service and tossed their horrible flatirons.

By "building an electrical consciousness," GE and other power conglomerates fueled Madison Avenue, which was keen on empowering women with new, labor-saving appliances (although these revolutionary devices did little to encourage men to help with cleaning, cooking, and other housework).

THE SATURDAY EVENING POST June 30, 1928

First aid in entertaining

FIRST aid in entertaining, nowadays, is a General Electric Refrigerator. If you have one, you can make those delicious, *different* salads and aspics and desserts. You can make them at your leisure—the same morning, or the day before, or whenever you wish. There's plenty of room to store them. There's the scientific cold to keep them fresh.

You will like the simplicity of the General Electric Refrigerator. It never needs oiling. It hasn't a drain-pipe or a belt or a fan. All its machinery is enclosed in the hermetically sealed casing mounted on top of the cabinet. And there is a remarkably large shelf area because the chilling chamber is so compact. All the models are up on legs, so the floor under them can easily be cleaned.

General Electric's world-famed engineers and scientists worked for fifteen years to develop this supremely practical and efficient refrigerator—to make it quiet, portable and economical to run.

Already it is giving flawless service in more than fifty thousand homes. Ask these users about the General Electric and make comparisons before you buy. There is a wide range of models and prices. Write us today for Booklet S-6B, which is completely descriptive.

GENERAL ELECTRIC
Refrigerator

ELECTRIC REFRIGERATION DEPARTMENT · OF GENERAL ELECTRIC COMPANY · HANNA BUILDING · CLEVELAND, OHIO

A 1920s General Electric advertisement pushes a new "supremely practical and efficient refrigerator" as essential "first aid" for entertaining guests.

Who would want to be without electricity when your neighbor was sure to acquire all these trappings of modern life? *Conspicuous consumption*, a phrase coined by sociologist Thorstein Veblen at the University of Chicago, became the sotto voce of a new age of keeping up with the Joneses.

On the Tesla Paper Trail

Although my government FOIAs on Tesla's papers weren't yielding anything, I received offers to speak about Tesla—first in a memorial lecture at a large, suburban Chicago library, then at the Serbian Cultural Center in Chicago. It was gratifying that the world wanted to know more about Tesla, but I wasn't sure that I could share anything new other than his remarkable boom-and-bust saga.

Yet like a phoenix, my fortunes improved. I was heartened to see that Tesla's name and brand value were about to be resurrected in a spectacular way by three Silicon Valley engineers who wanted to build a superior electric car named for the inventor. When the car company offered its stock to the public in 2010, shepherded by PayPal co-founder Elon Musk, a new wave of interest in Tesla, the inventor, swept the world.

The following year something in the quantum intelligence of the ether connected me to someone I needed to talk to: Vladimir Jelenkovic, then director of the Tesla Museum in Belgrade, Serbia. I received a call from a public relations person, saying that Jelenkovic would be in Chicago to talk about a small touring Tesla exhibit in the summer of 2011 that had only one other destination on the planet: Perth, Australia.

Chicago was picked as the only North American site due to its large Serbian population, many of whom had emigrated to the South Side and beyond to work in the steel mills and meatpacking plants. (Perth is also home to a large Serbian community.) I had seen a slice of Serbo-Croatian America when I was starting my career, working in South Chicago as a cub reporter for the *Daily Calumet*. I would marvel at the Serbian Orthodox church and local haunts like the Golden Shell, a restaurant that featured belly dancers, Balkan cuisine, and the toasting drink of choice: slivovitz, a lovely plum brandy.

In Jelenkovic, I found a kindred spirit: a journalist by training who had covered science and had been on television. He had founded the science magazine *SciTech* and represented the software company Oracle in Nigeria and Serbia. Above all, Jelenkovic was a world-class Tesla promoter. With his tiny traveling exhibit, Jelenkovic was attempting to bring Tesla back into the mainstream after nearly a century of exile. Tucked away in an obscure, rarely used section of Chicago's Navy Pier, the display featured Tesla's Eggs of Columbus and basic accomplishments of the inventor.

When I had lunch with Jelenkovic, he opened the door on my constricted knowledge of the documentation of Tesla's business relationships with Morgan and others. Not only did he have letters between Tesla and Morgan in his collection (which I later acquired from the Morgan library); he possessed more than 100 pages of letters between Tesla and Insull. Since I had spent hundreds of hours combing through four different archives looking for Insull correspondence, including the collections of Edison, FDR, and Loyola University, this was an exciting development. During the course of my research into Insull, I believed I had discovered everything about the man. This new letter series was a genuine breakthrough

Regaling me with the short history of Tesla's achievements and decline, Jelenkovic was more than a historian; he was a global advocate for the rehabilitation of Tesla's reputation as a visionary humanitarian inventor. Like most Teslaphiles, Jelenkovic disparaged Edison's achievements. "Edison's 100-watt lightbulb is now *banned* in Europe," Jelenkovic noted with disdain. "Some 70 percent of its output is heat. AC, on the other hand, is used by all large power systems."

After a delightful lunch, I was eager to ask Jelenkovic what had fed my interest in Tesla after reading the 1935 letter to Insull: Did he have any documents with detailed plans on the "death ray" or anything else that emerged from the Wardenclyffe project?

"We have all of Tesla's legacy, but not one letter about the death ray. Don't you have them?"

Although my meetings with Jelenkovic, Terbo, and a host of Tesla luminaries at the Hotel New Yorker in 2012 restored my faith in the global revival of Tesla's importance, I was still striking out on the search for Tesla's lost files.

Gene Morris, an archivist at the National Archives and Record Administration, kindly wrote me a two-page letter explaining that the National Archives didn't have *any* Tesla records "within our custody." Anything held by the Office of Alien Property, the temporary World War II agency, had been "turned over to Sava Kosanovich, Mr. Tesla's heir, a few years after the end of the Second World War." Kosanovich had then sent Tesla's papers to the Tesla Museum in Belgrade, although it was never clear what exactly Kosanovich had possessed and what the government kept for itself or simply copied.

What remained in the sketchy FBI memoranda were notes from Kosanovich, then with the Serbian consulate in 1943, working with Tesla's friend Kenneth Swezey to remove a book of Tesla testimonials and three pictures of the inventor from a safe in the hotel room. Yet the OAP carted off "two truckloads and all of the property of Tesla" to a Manhattan warehouse, where the material was placed alongside "thirty barrels and bundles belonging to Tesla, which had been there since 1934."

What was the government doing with Tesla's documents prior to his death in 1943? Again, the FBI doesn't say, although the author of the redacted 1943 memo I saw notes that "jurisdiction over this property is doubtful."

Needless to say, if you're a Tesla conspiracy theorist, this kind of limited information will drive you crazy. In another memo, dated January 20, 1951, Hoover's assistant and confidant (and perhaps lover), Clyde Tolson, cites the interest of Abraham Spanel, "who had contacted the War Department the day before Tesla died to make available certain patents." Then the memo ends, maddeningly without any further explanation.

Was Spanel interested in Tesla's particle beam ("death ray") drawings—if they existed at all? Or was he searching for schematics on how telegeodynamics worked? While many people have tried to draw connections between Tesla, the FBI, and the Department of Defense as well as Soviet Union research on particle weapons, no official conclusions as to if or how Tesla's material was used have emerged over time.

In the late 1970s, the Pentagon was shocked when the Soviet Union claimed it was working on particle beam weapons. Then came President Ronald Reagan and his "Star Wars" weapon program to shoot down incoming

Clyde Tolson, shown here with his boss (and possible lover) J. Edgar Hoover at a boxing match in 1936, claimed that International Latex Corporation founder Abraham Spanel contacted the War Department the day before Tesla died, requesting that it release certain patents from its custody.

intercontinental ballistic missiles. The program never produced an effective weapon—as far as we know—and was canceled to little fanfare.

Yet I was still intrigued by Tesla's telegeodynamics proposal, which seemed much more plausible. The earth can conduct electricity because it has a nickel-iron core. It also transmits tremendous amounts of mechanical energy as its tectonic plates shift, causing earthquakes from California to the Mediterranean. The more plausible application is that electricity could be delivered anywhere in the world without investing trillions in infrastructure. Could excess power from the hydroelectric-rich Pacific Northwest or Tennessee Valley be sent through the earth to Borneo or India or Africa? Could remote villages along the Amazon or in the Sahara be lit up? Sub-Saharan Africa could emerge from the darkness, as would Bangladesh and inner Mongolia.

There's no conclusive science behind telegeodynamics as this book goes to press. As noted in his book *Tesla: Inventor of the Electric Age*, W. Bernard Carlson is skeptical that such a technology would work on a large scale. Modern geologists are still struggling with earthquake prediction; using the earth as an energy conduit is not high on their priority list. But grand theories were Tesla's stock in trade, and their possibilities still intrigue scientists and engineers.

In a letter to Tesla from Westinghouse engineering director M.W. Smith on November 18, 1940, the executive refers to Tesla's "art of producing terrestrial motions at a distance." The letter, though, was yet another polite rebuff to the inventor on his telegeodynamics proposal. The following year, Tesla wrote a rambling letter to Westinghouse about spending billions of dollars on a growth factor for chickens that would produce better meat and

Electrical Experimenter

233 FULTON STREET, NEW YORK

Publisht by Experimenter Publishing Company, Inc. (H. Gernsback, President; S. Gernsback, Treasurer;) 233 Fulton Street, New York

Vol. VI Whole No. 70 FEBRUARY, 1919 No. 10

NIKOLA TESLA AND HIS WIRELESS LIGHT............Front Cover
From a painting by George Wall.
TIDAL POWER PROBLEM SOLVED AT LAST.................. 685
PRODUCING RAIN BY ELECTRICITY AND X-RAYS............ 687
MOVING PLATFORM FOR NEW YORK'S CROSSTOWN SUBWAY..............................By H. Winfield Secor 688
THE UNKNOWN PURPLE........................By Dorothy Kent 690
ZEPPELIN FLIES FROM BULGARIA TO KHARTOUM AND RETURN WITHOUT STOPPING.................................. 691
FAMOUS SCIENTIFIC ILLUSIONS..........By Dr. Nikola Tesla 692
SOLDIERS' ILLS CURED BY ELECTRICITY..................... 695
MY INVENTIONS—NO. 1 OF A SERIES
By Dr. Nikola Tesla. Exclusive Feature. 696
SUBWAYS OF DOWN-TOWN NEW YORK
From a painting by George Wall. 699
POPULAR ASTRONOMY—THE MILKY WAY
By Isabel M. Lewis, of the U. S. Naval Observatory 700
WOMEN NOW TRAINED AS METER READERS................ 702
LARGEST ELECTRIC CRANE LIFTS COMPLETE TUG BOAT... 703
SELLING ELECTRICITY BY THE "CAN......................... 704
LATEST ELECTRICAL APPARATUS AND NOVELTIES........ 705
WIRELESS LEGISLATION NEWS.............................. 706
RADIO DEPARTMENT—PRESIDENT WILSON IN TOUCH WITH AMERICA ON LAND OR SEA—VIA RADIO.................. 708
VACUUM VALVE ACTION AND THE ELECTRIC CURRENT
By K. G. Ormiston, Instructor in Radio 710
THE VORTEX RING THEORY OF THE ELECTRON
By F. W. Russell and J. L. Clifford 712
"BALL LIGHTNING" EXPERIMENTS..By Samuel S. Weisiger, Jr. 714
A USEFUL ELECTRICAL LABORATORY SWITCH-BOARD
By H. Danner 715
EXPERIMENTS IN RADIO-ACTIVITY
By Ivan Crawford. Part II.—Ionization 716
EXPERIMENTAL MECHANICS—LESSON X..By Samuel D. Cohen 717
EXPERIMENTAL CHEMISTRY—THIRTY-THIRD LESSON
By Albert W. Wilsdon 718
HOW-TO-MAKE-IT DEPARTMENT—PRIZE CONTEST.......... 719
WRINKLES, RECIPES AND FORMULAS...Edited by S. Gernsback 720
LATEST PATENTS DIGEST..................................... 721
WITH THE AMATEURS—LABORATORY PHOTO CONTEST.... 722
PHONEY PATENT CONTEST..................................... 723

EDITORIAL

The New Wireless

IT will come as a profound shock to all wireless enthusiasts, scientific and amateur alike, that their present-day notions on wireless are totally erroneous and not based upon actual facts. For years we clung to the theory that a wireless message radiates from the aerial wires of the sending station and speeds over the surface of the earth thru the ether towards the receiving station. We thought that we were sending out pure Hertzian waves from our transmitters. We thought that we received these waves over the aerial wires of our receiving station. All of these theories are wrong and will be relegated shortly into the past along with the early notion that the earth stood still, while sun, moon and stars revolved around it.

Remain only the physical facts that we did send and did receive messages without wires—but they are not sent by means of pure Hertz waves, nor do they go by way of the ether as radiations.

In a highly illuminating article printed elsewhere in this issue, Nikola Tesla explodes all of our present orthodox views as to wireless propagation and makes it clear that the earth is the sole medium thru which our wireless impulses travel, in the form of true conduction. Particularly does this hold true for long distance messages: Here we are sending out a compound impulse three quarters of which is a galvanic current, traveling thru the conducting earth, the other quarter or less is in the form of Hertz waves, going by way of the ether. This explains why we can send signals to airplanes and vice versa; but even here we probably have to do not with pure Hertz waves: it is almost certain that we have capacity-inductive effects as well.

Tesla maintaining that there can be no long distance effects by radiations transmitted thru the ether, but rather only by currents thru the earth, it follows that in his opinion all our radio apparatus is designed and operated faultily. Indeed, this is not a brand new idea of the famous inventor. He has been preaching it ever since he took out his first patents and described his system in 1893—long before Marconi thought of wireless. But he was preaching to a stone deaf scientific world.

But how simple it all becomes when we stop to apply a little reason and logic to Tesla's claims. For instance, we can send radio impulses three to five times as far over salt water as over land. Why? Simply because the impulses go *thru* the water, which is a much better conductor than earth alone. If we were sending pure Hertzian waves, why do we connect one wire at both sending and receiving station to the ground? Hertz never dreamt of such a thing. If you are still unconvinced that the earth is the chief medium of transmission, disconnect your ground wires entirely and try to send and receive. Now you may work with Hertz waves, but the distances you can bridge will be pitifully small.

Already Tesla's logic is filtering into our radio scientists' minds. *All the big stations are beginning to scrap their towers and aerial wires, at least for receiving. They now bury their "aerial" wires in the ground*, and lo! they can receive signals twice as far as before. Incredible, but it is being done every day. And—wonders upon wonders—how we will laugh at our present and past blindness—*the static interference is practically gone* the minute we pull our aerial wires down and bury them! Static Electricity? There never was a reason for having the bugaboo, for there is no "static" in the ground.

But Tesla goes much farther. In time he will show the world wireless power transmission effected *not by ether waves but by currents thru the earth*, which is a first rate conductor. Like all big things, the problem is simple. At some point on the globe he will erect a station powerful enough to charge the whole earth with electricity—and keep it charged. To do this we need about 10,000 kilowatts. Then at *any* point on the globe the current can be tapt by means of suitable apparatus. Like a bell ringing transformer, *connected* to your supply line, no current is consumed unless you close the secondary circuit. Tesla's world wireless works just that way. *No current is consumed till it is tapt at the distant receiving station.*

H. GERNSBACK.

The ELECTRICAL EXPERIMENTER is publisht on the 15th of each month at 233 Fulton Street, New York. There are 12 numbers per year. Subscription price is $2.00 a year in U. S. and possessions. Canada and foreign countries, $2.50 a year. U. S. coin as well as U. S. stamps accepted (no foreign coins or stamps). Single copies, 20 cents each. A sample copy will be sent gratis on request. Checks and money orders should be drawn to order of EXPERIMENTER PUBLISHING CO., INC. If you change your address notify us promptly, in order that copies are not miscarried or lost. A green wrapper indicates expiration. No copies sent after expiration.

All communications and contributions to this journal should be address to: Editor, ELECTRICAL EXPERIMENTER, 233 Fulton Street, New York. Unaccepted contributions cannot be returned unless full postage has been included. ALL accepted contributions are paid for on publication. A special rate is paid for novel experiments; good photographs accompanying them are highly desirable.

ELECTRICAL EXPERIMENTER. Monthly. Entered as second-class matter at the New York Post Office under Act of Congress of March 3, 1879. Title registered U. S. Patent Office. Copyright, 1918, by E. P. Co., Inc., New York. The Contents of this magazine are copyrighted and must not be reproduced without giving full credit to the publication.

The ELECTRICAL EXPERIMENTER is for sale at all newsstands in the United States and Canada; also at Brentano's, 37 Avenue de l'Opera, Paris.

682

In this *Electrical Experimenter* editorial from 1919, publisher Hugo Gernsback makes a passionate case for Tesla's telegeodynamics theory, noting that current wireless communication consists of "a compound impulse three-quarters of which is a galvanic current, traveling thru the conducting earth, the other quarter or less [of which] is in the form of Hertz waves, going by way of the ether [air]." He points out that radio stations are beginning to bury their receivers in the ground and "can receive signals twice as far as before."

eggs. I did not find a reply to the letter, which may have been his last correspondence with Westinghouse.

Unfortunately, the Office of Information Policy at the Department of Justice has thus far denied my administrative appeal to reveal the blacked-out names in the FBI memos. Meanwhile, all but the Defense Advanced Research Projects Agency (DARPA) have failed to respond to my Department of Defense FOIA requests. What remains of the complete corpus of Tesla documents is either in Belgrade or some U.S. government warehouse—or both. It's known that the anticommunist witch-hunter Senator Joseph McCarthy took an interest in the Tesla papers and their journey to (then communist-controlled) Belgrade. A military man named Ralph Bergstrasser, who knew Tesla and may have copied his papers, was the one who alerted McCarthy and the FBI to their existence with a thirty-page letter in the early 1950s.

Are all of these facts connected? Perhaps in a loose way. Tesla's ideas *inspired* weapons, but did he provide detailed, working engineering schematics to actually *build* them using the technology of his day?

As noted earlier, Tesla believed his device could vastly exceed the speed of light, which is impossible according to Einstein's relativity theory. Regardless, an energy source far beyond what was available at the time would be required to produce the kind of power Tesla proposed. Ultrapowerful particle accelerators would come decades after his ambitious proposal. It takes trillions of electron volts to produce highly controversial basic "God" particles like the Higgs boson, and that massive quantity of energy was only possible using technology that became available in recent years, through the contribution of billions of dollars by several countries. Perhaps tapping the sun's thermonuclear capabilities in a fusion reactor would do the trick, but that technology has proved elusive after decades and billions of dollars in global research. Nevertheless, this colossal stellar energy is akin to a technical Holy Grail that could provide vast quantities of carbon-free energy.

Marc Seifer, who has done the lion's share of Tesla research over the years and based his densely detailed 1987 doctoral dissertation on the inventor's work, found evidence that Tesla *did* openly offer his weapon ideas to the U.S. government, so the idea of a government "conspiracy" to "steal" his

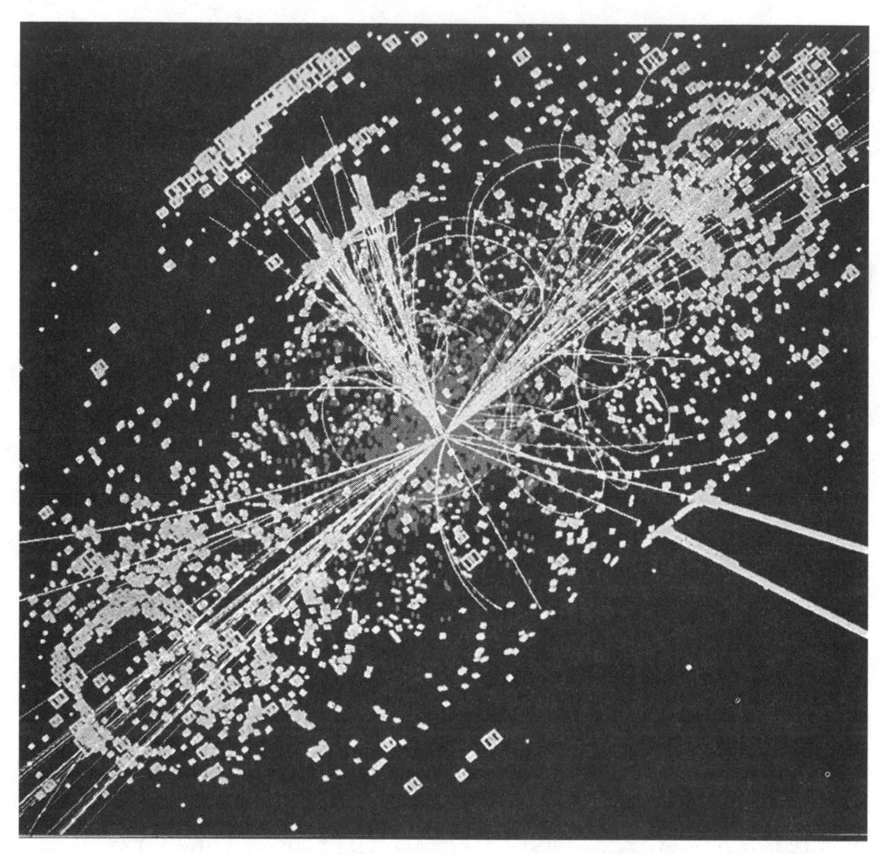

An example of simulated data modeled for the CMS particle detector on the Large Hadron Collider (LHC) at CERN. Here, following a collision of two protons, a Higgs boson is produced that decays into two jets of hadrons and two electrons. The lines represent the possible paths of particles produced by the proton-proton collision in the detector, while the tiny rectangular fragments depict the energy deposit.

ideas is probably false. "I wouldn't say they were 'stolen,'" Seifer told me in an email as I was completing my research. "Tesla was working with the War Department and offered it to them [the government]. I think they tried to develop it. They kept it top secret, as was his patent; it was never published as it was in that secret category."

"The main secret weapon paper was the Particle Beam paper," Seifer added. "There may be stuff on cosmic rays or even tachyonic particles, but I doubt it. I know there was a key bladeless turbine paper that no one seems to be able to locate. Ultimately, I think we are talking about Tesla's particle beam secret patent." Seifer noted that his seminal biography of Tesla, *Wizard: The Life and Times of Nikola Tesla,* gives "the real story": "That [particle beam] paper was squirreled away to Wright Patterson Airforce base. But at the same time, they left a copy with Tesla's effects and the [original] paper is at the Museum in Belgrade," Seifer asserted.

Did the secret paper result in a viable weapon thirty years ago? It doesn't seem likely, although both the United States and the Soviet Union probably spent billions on the concept. More than two decades later, both the Soviet Union and the United States were said to be attempting to develop particle beam weapons (and probably still are).

"The Russians obviously had access to it, as they published a schematic of the particle beam weapon (reproduced in *Aviation Week* in the 1970s) seven or so years before Tesla's secret patent was published," Seifer told me. "Since Yugoslavia was behind the Iron Curtain [at the time], the Russians had access to Tesla's papers at the Museum and they probably saw it first in the 1950s."

Is such an awful weapon *possible?* I'm not an engineer, but with the advances in high-energy physics and perhaps power through nuclear fusion reactors (still in development), who knows?

Be Empathic

As I noted earlier, Tesla insisted that his celibacy was akin to being married to his work. What's probably not disputed is that Tesla was probably very lonely in the last decades of his life, particularly when his financial backers abandoned him. Westinghouse, J.P. Morgan, and Astor had all died when he began his nocturnal journeys through New York City.

A new school of neuroscience is finding that we are happiest and perhaps most productive when we are *empathically* among and thinking of other people. University of Chicago neuroscientist John Cacioppo has found that magnetic resonance imaging (MRI) scans of the brain show increased activities of higher-level thinking when we are in an empathy mode. Being lonely or isolated, Cacioppo and his colleagues discovered, was correlated with "decreased executive function," which is involved in complex thinking and decisions. Those who described themselves as lonely also tended to have shorter lives than those actively engaged in social relationships.

"The social environment is profoundly important," Dr. Cacioppo says. "When we feel connected, we are generally less agitated and less stressed than when we feel lonely.... All of which can have profoundly positive influences on our health."

Having a partner or being married can prolong life, studies show. It's difficult for us to fully reap all the benefits of life in isolation from the broader world. We need social sounding boards. It's not clear whether loneliness enhances or stunts creativity, but in Tesla's case—even though he was in fairly good health well into his ninth decade—it's clear that he was not as productive or successful as he had been while in his dynamic social circle.

When I visited Tesla's rooms in the Hotel New Yorker, I couldn't get over how small they were. The image of Tesla and his pigeons in this tiny space was distressing to me. What if he had been engaged with young people or holding court with other engineers? What would his last year have been like if he had communed or collaborated with the great physicists of the day like Werner Heisenberg, Niels Bohr, Wolfgang Pauli, or Enrico Fermi?

Working together is one of the keys to taking genius—your own and others'—to the next level. When I asked Walter Isaacson, president and CEO of the Aspen Institute (a global nonpartisan, nonprofit organization for education, leadership, and policy studies), what new subject should be taught in schools to foster innovation, he immediately shot back: "Collaboration." A world-class journalist and biographer of Ben Franklin, Albert Einstein, Henry Kissinger, and Steve Jobs, Isaacson has documented the world-changing power of idea-sharing and teamwork among the brightest minds in recent history.

So what can we learn from examinations into the effects of isolation and collaboration? To maximize creativity:

- **Become involved in your family and community.** What needs to be done? Where can you apply your skills and time to best help others? When I turned fifty, I started a nonprofit dedicated to helping taxpayers. Join civic groups to volunteer and inspire others.

- **Choose engagement over isolation.** Younger people have energy and ideas; older people have experience and wisdom. It's not always a perfect marriage, but these qualities are complementary and have the potential to mend our tattered social fabric. Don't hibernate; collaborate.

- **Be a lifelong learner.** Tesla learned and used several languages. He was just as at home in Paris, Budapest, and the Balkans as he was in the heart of New York City. Embrace travel, meet people from an array of cultural backgrounds, and be open to others' ideas.

To paraphrase the poet John Donne, no one is an island. Changing the world requires a raucous group of collaborators with different ideas and skill sets who may not always agree with us. Truly disruptive innovation often comes from diversity of ideas, not lonely inspiration.

X

THE ONCE AND FUTURE TESLA

THE WIZARD'S TWENTY-FIRST-CENTURY LEGACY

There is I believe a reason for the bizarre fact that it was Edison and not Tesla that became and still is the icon of 'genius inventor': Edison was a character that fitted snugly into the zeitgeist of the twentieth century; Tesla belongs to the twenty-first.

—Dino Karabeg, Professor, University of Oslo

Philadelphia's Independence Mall was pulsating with energy on the eve of the 159th anniversary of Tesla's birth. On that perfect July night, Teslaphiles had gathered from all over the world, not only to celebrate the inventor's birthday at midnight but to observe the spirit of the once and future Tesla.

Along the length of the mall, which is a block-wide national park, sits the iconic Independence Hall, where the Declaration of Independence was signed and where Franklin supposedly quipped, "We must, indeed, all hang together, or most assuredly we shall all hang separately." A few blocks to the east, Franklin had built his commodious house, installing some of the world's first lightning rods. The original house was razed in 1812; its footprint is outlined by a steel "ghost structure" designed by architect Robert Venturi. A museum occupies the present site. Rounding out the Independence Mall

Nikola Tesla passes 500,000 volts of electric current through his body to power a fluorescent wireless light at his Houston Street laboratory in New York City, 1898.

complex are the National Constitution Center and hall where the American Philosophical Society is headquartered.

Nick Lonchar, the able torchbearer for all things Tesla, was leading the birthday celebration, which featured a Tesla coil demonstration, a few speeches, and testimonials by the diverse group gathered on the east side of the building housing the Liberty Bell. The event was sponsored by the Philadelphia-based Tesla Science Foundation, which Lonchar founded "as a club for inventors." The group has created a traveling Tesla exhibit and placed busts of Tesla in various key places, including the Franklin Institute. The foundation has also been attempting to establish a permanent space for a working laboratory and Tesla museum while collaborating with artists to create exquisite installations and music to honor the inventor.

Philadelphia's iconic Independence Hall was the backdrop for the "Energy Independence Celebration," which coincided with the 159th anniversary of Tesla's birth on July 10, 2015.

Lonchar, who was named after Tesla, is a private detective and locksmith by trade who grew up in Belgrade. He has his own theories about Tesla's last days and the fate of the lost papers, and he's still investigating what transpired after the death of his hero.

"Tesla is the father of the wireless age," Lonchar told me. "It's *his* time." But according to these enthusiastic Tesla fans, Tesla is not merely an innovative scientist. To Lonchar—and many Serbians—Tesla is literally a saint who deserves official spiritual recognition; other Teslaphiles have started a petition to have the Serbian Orthodox Church canonize the inventor. "We view Tesla as a saint of science and knowledge and wish to present him that way," Lonchar says. "Tesla was a man who wanted life to be better for all people."

Like a virtuoso jazz improviser, Lonchar kept the energy flowing at Independence Mall. The event was a mash-up of the entire spectrum of Tesla ideas and lore as well as an "Energy Independence Celebration." No one at the event was enamored of fossil fuels or other current methods of producing and distributing energy. On the other side of the park, closer to Independence Hall, DJs were blasting electronic music into the gentle summer night as people wandered in from the street to dance and check out the handful of Tesla exhibits.

When the electricity powering the lights and Tesla coil mysteriously went out, I had a chance to commune with the diverse array of Teslaphiles at the event. Sherry Kumar, also an officer with the foundation who does public relations out of New York, welcomed those gathered, passing around a card embossed with the "Declaration of Energy Independence" that was drafted by "the free people of earth" in 2009. It's a manifesto for sustainable energy.

Sam Mason, who had been demonstrating a Tesla coil and illuminating a wireless fluorescent bulb, is a director of the Tesla Science Foundation and an omnipresent guide at Tesla events. This time I had a chance to talk with him about his work on antigravity propulsion for future spacecraft, which has its origins in Tesla's research. Upon reading Tesla's 1890s research on high-voltage discharges, a scientist named T. Townsend Brown discovered that electrostatic and gravitational fields may be closely related. If this technology, which has been explored by NASA, proves technically feasible, it could offer a novel means of rocket propulsion using "asymmetric capacitors" and electromagnetic fields or "electrogravitics" to move a spaceship. Mason is working on developing a generator that would create such a force. Not fully understanding his engineering jargon, I went straight to the question that most skeptics ask of this speculative but promising technology: Will it work? Could Tesla's flying machine get off the ground with wireless power? "Innovation has never been a straight line," Mason replied. "The best ideas come to me on the way to the Jersey shore."

After the Independence Hall clock tolled midnight, a Tesla birthday cake was cut and distributed. Everyone toasted the inventor and a wireless, sustainable future.

The following morning, the Tesla's People's Conference was held (the first half was mostly in Serbian), followed by a wider presentation for English speakers in the Ethical Society of Philadelphia hall off charming Rittenhouse Square. The small, elegant park was bustling with jazz musicians, children, the elderly, and strollers from every walk of society.

At the conference, a broad range of Tesla advocates (including myself) spoke briefly. Mason brought the group up to date on Tesla conferences in Turkey and Belgrade earlier in the year. Joe Kinney described the influx of Teslaphiles into the Hotel New Yorker. Educator Ashley Redfearn explained how she's integrating Tesla into her school's science education program. Author Mark Passio wildly expounded on why he thinks one of Tesla's telegeodynamic devices brought down the World Trade Center on September 11, 2001. Sherry Kumar gave an update on a project to run wireless electric buses. Thereminist Mano Divina and his Divine Hand Ensemble played some lovely music. Serbian filmmaker Zeljko Mirkovic filmed a documentary of the event.

Philadelphia-based artist and educator Brian Yetzer explained how he had designed the Nikola Tesla "augmented reality" app alongside brilliant images for the traveling Tesla exhibit. These colorful images (featured in the middle of this book) trigger interactive content on mobile devices, delivering instant educational videos about Tesla and his technology. "Technology is seen as magic before it's adopted by the masses," Yetzer observed. As his phone scanned his poster of a Tesla coil, the screen brought it to life, displaying a functional 3-D animation explaining the device. "Wouldn't it be cool to know how a Tesla coil works?"

Other conference attendees described how they are incorporating the idealism of free energy from the earth and sky into everyday living. Rodney Leatherman, a Tennessee entrepreneur, is building green homes in the Smoky Mountains that will require "zero energy" (that is, they provide their own power). "These communities will produce their own power from 750-square-foot homes with wind turbines and geothermal heating," Leatherman explained.

Around us, the city of peace, revolution, and invention was wobbling its way deeper into the twenty-first century. Ideas floated around Rittenhouse Square like kites about to get stuck in trees rooted in our everyday reality. The port town carries this spirit well, wearing her dowager's shawl with optimism for the future.

Saving Tesla's Dream Palace

Did it matter if Tesla's most outré ideas worked as viable technologies? That was the piercing question I faced as I changed course from unearthing his lost papers to focusing on his many luminous endowments to humanity.

I decided to visit rooms 3327 and 3328 of the Hotel New Yorker, where the inventor died. Towering above Eighth Avenue, just blocks from the *New York Times* building, the Port Authority Terminal, Madison Square Garden, and the Empire State Building, this Art Deco masterpiece embodies much of what Tesla had worked for in his professional life. When you enter its spacious, bustling lobby, you can imagine the hotel trumpeting the virtues of the modern world in 1930 as its mammoth cousin, the Empire State Building, was rising a few blocks to the east.

Currently owned by the Reverend Sam Myung Moon's Unification Church and managed by the Wyndham Hotel chain, the Hotel New Yorker is a Lourdes for tech mavens, mostly because of Tesla. The forty-three-story skyscraper was the tallest hotel in New York when it was finished in 1930, and it was one of the first self-sufficient hotel "cities within a city" to be built anywhere. Tesla would have felt at home when he moved there in 1933.

In the basement, you see why the New Yorker became a beacon for modernity, particularly if you take the popular tour offered by Joe Kinney, the building's chief engineer and unofficial archivist. The hotel once had an operating generator to provide enough power to support 35,000 people. Although the hotel has long been on the electrical grid, Kinney showed me the long-idle, massive generator and the infrastructure around it. The hotel's other working parts were no less impressive: a kitchen that occupied an entire acre, a hospital with an operating room, a radio broadcasting station, and an indoor skating rink. All this was possible because the hotel's generators could supply generous amounts of power. There was even a tunnel (now sealed off) that led directly to nearby Penn Station.

Kinney mentioned that "he didn't know if Tesla ever came down here," but he added with a puckish smile that the inventor probably knew about how his contribution had created such an enterprise.

The massive electrical generator in the basement of the Hotel New Yorker once supplied enough power for 35,000 people.

I wandered up to Tesla's suite, which was actually composed of two rooms with memorial plaques on each door. Since the hotel allows people to stay in Tesla's suite, I chose not to knock on the doors, though. Over the years, a stream of officials from the Balkans and parts beyond has visited the hotel, which has become ground zero for nearly everything connected with Tesla. But there was still Wardenclyffe, decaying on the east end of Long Island like the lone, decapitated monument in Percy Shelley's sonnet "Ozymandias."

My first New York City Tesla conference in 2012 came as a revelation. William Terbo, New York–based Tesla Science Center president Jane Alcorn, (then) Tesla Museum director Vlad Jelenkovic, and Tesla savant Marc Seifer all spoke. This was the World Series for Teslaphiles, highlighting his many achievements and focusing on how to best preserve his legacy.

Jelenkovic led off the seminar with the invocation that "Tesla remains one of the most intriguing minds of all time." As a bonus, he had prepared a computer-animated re-creation of what Tesla's Wardenclyffe lab may have looked like in its heyday. A virtual tour walked viewers through the lab. The audience was thrilled.

Joe Kinney of the Hotel New Yorker and Aleksandar Protic, a UNESCO official and director of the Tesla Memory Project, followed Jelenkovic. Decades in the making, the global effort to recognize and build upon Tesla's achievements was growing.

The grand finale of the conference, one reminiscent of a Broadway closing number, was the appearance of Matt Inman, a Seattle cartoonist who produces the satirical comic *The Oatmeal*. Like many other Teslaphiles, the author of humorous books such as *How to Tell If Your Cat Is Planning to Kill You* felt a sense of injustice that Tesla's reputation had been overshadowed by Edison. When his comic comparison "Tesla v. Edison" scored more than 400,000 likes (spoiler alert: Edison came out looking evil), Inman knew he had tapped an underground sentiment of modern loyalty to the Serbian inventor. "He suffered for his work," Inman quietly noted. When Inman then told of his crowd-funding campaign to purchase the property where the abandoned Wardenclyffe laboratory stands—an effort that ultimately raised $1.4 million from more than 30,000 donors—the crowd erupted in applause and gave him a standing ovation.

When I came through Long Island toward the end of this century's first decade, the Wardenclyffe site lay in ruins. Actually, it was worse than that: One of its recent occupants, the Peerless Photo Corporation, had dumped an array of toxic chemicals in and around the site. The majestic building with its crown-shaped chimney cap was deteriorating by the day. Like Tesla in the early twentieth century, Wardenclyffe needed a financial savior, although the odds of the dilapidated complex finding anyone with deep pockets willing to spend tens of millions of dollars on rehabilitation appeared slim. After all, this was at the height of the worst recession since the 1930s. The avaricious moguls who had their Hampton palaces just a few miles east of the Shoreham site were selling their Gatsby dreams then, just trying to keep their heads above water. But thanks to Inman's 2012 crowd-funding campaign and aided by a $1 million pledge by Elon Musk, the Shoreham property was purchased and the project to extend Tesla's legacy into a working education center at Wardenclyffe was on its way. All told, though, Alcorn said it would take at least $10 million to get the Long Island site in shape for visitors—an ongoing project that will take years but eventually introduce thousands to Tesla's astonishing work and vision.

Tesla's Biggest Twenty-First-Century Champion

When I asked William Terbo about the billionaire visionary Elon Musk, Tesla's grandnephew said he was grateful for the entrepreneur's support but noted "he owes me a car."

In his 1919 autobiography, Nikola Tesla made the following boast:

> “As early as 1898 I proposed to representatives of a large manufacturing concern the construction and public exhibition of an automobile carriage which, left to itself, would perform a great variety of operations involving something akin to judgment. But my proposal was deemed chimerical at that time and nothing came from it.”

If there was ever a more potent global recognition of the Tesla narrative, it came in spades 110 years after his initial proposal with the launching of the Tesla Roadster in 2008. While it certainly wasn't the first all-electric car, it was one of the sleekest, sexiest, and fastest cars without a tailpipe. It also revived Tesla's legacy, this time as a hot global brand and potent status symbol for the wealthy.

Conceived in 2003 by Silicon Valley engineers Marc Tarpenning and Martin Eberhard, Tesla Motors is financed by Elon Musk and other venture capitalists. At the start of the twenty-first century, the Toyota® Prius—a hybrid that successfully combined an electric motor with a gasoline-powered engine (we own one)—was becoming the consumer badge of progress for the nation's movie stars and elite. The next step was an auto that had no tailpipe emissions at all. Tarpenning and Eberhard, needing $7 million to build a prototype, approached Musk, who ponied up $6.5 million, becoming the company's largest investor.

Flush with cash from his PayPal investment, Musk was working on a much more ambitious venture: SpaceX. The entrepreneur wanted to launch reusable rockets in space—multistage rockets whose components could be landed, intact, back on earth or on floating ocean pads! Such a project would not only save billions over time; it would also enable private companies to launch their own satellites and space exploration missions. The idea was to mass-produce these rockets so the per-unit cost would drop, and as of this writing, there have been four successful vertical rocket first-stage (booster) landings following a return from low earth orbit.

Meanwhile, J.B. Straubel, who went to Stanford to study physics and lived about a half-mile from the Tesla offices, wanted to power a car with lithium ion batteries, which were then (and now) used to power cellphones and other

SpaceX's Falcon 9 rocket launched the ORBCOMM™-OG2 Mission 1 on July 14, 2014. It successfully deployed six telecommunication satellites.

small devices. At his first luncheon meeting with Musk, he was reportedly offered $10,000 to oversee the vehicles' technical and engineering design.

Conceptually, the Tesla is a luxurious body embedded with a motor, computer brain, wheels, and batteries. But the engineering soon began to take on a new life of its own as the designers discovered they needed a lot of juice to power a high-performance car with a big price tag. The original technical spec for 20 lithium batteries grew to nearly 7,000. And there was the recurring danger of a fire, particularly if the batteries were exposed to air. The last thing the engineers and investors wanted, after investing tens of millions of dollars, was an overpriced exploding car.

After some internal management disputes, Eberhard was removed as CEO as the company bled money during the car's development. Musk aggressively promoted a car that had an ever-shifting initial sales date and ever-escalating price. His SpaceX test launches were like the early days of the space race—a lot of publicity followed by rocket explosions. The six-figure Roadster, though, proved a success and was followed in 2012 by the more pedestrian Model S sedan, which was listed for about $70,000 before tax credits. And orders are pouring in for the newer, even lower-priced Model 3.

Shortly after my Hotel New Yorker conference, my father expressed interest in the Model S Tesla. At age eighty-seven, he had safely outgrown his minivan and wanted something more elegant than the 1980s Mercury with a landau (fake convertible) roof he kept in Florida. My wife, Kathleen, agreed to take him for a test drive on a lovely summer day. The Tesla sedan, silent as a puma on the prowl, rode smoothly. And my father went back to his Toyota minivan with a sly smile, having achieved the bucket-list satisfaction of sitting in the most modern auto in the world.

What you don't see when you drive the Tesla S is the system behind it. Although you are the driver, it is controlled by a sophisticated computer that constantly monitors performance. Moreover, the Tesla car is connected to a much larger system that can install software updates over the Internet. It isn't just an electric car; it is a cybernetic semi-automaton, a four-wheel computer connected to engineers through infinite bandwidth. If I didn't have two girls to send to college and if I didn't want to be eating cat food in

This 2014 photograph shows the Tesla S in front of a newly opened showroom in Paris, France.

my retirement, I would have been more than eager to purchase this state-of-the-art vehicle for myself.

Having emerged as one of the sine qua non status symbols on the road, the Tesla Motors brand is a machine that the inventor would have embraced. New commands and upgrades can be sent to the car's brain wirelessly. If the car didn't need half a ton of batteries and could be recharged by energy in the atmosphere, it would be the absolute perfect car. As this book goes to press, the company is developing a version that could be eventually driven without a human behind the wheel. Eventually the Model S would exceed the venerable *Consumer Reports* rating system.

Nikola Tesla would have loved the Tesla car for another reason: It embodies a soon-to-be independently controlled unit, supported by a larger system. There are engineers downloading code to make the car run better from remote locations. And there is a growing network of ultrafast charging stations (charge-ups in minutes) so that owners won't suffer from "range anxiety" (that is, running out of charge) anyplace in the United States.

In an even greater leap of systems integration, Tesla is also selling independent battery arrays for your home. That means you can charge up Tesla batteries through the grid or roof-based solar panels (sold by another Musk company, Solar City). Although this is the opposite of Nikola Tesla's AC grid model—you produce your own DC power instead of relying on a big utility—it's a profound advance in future energy independence. Musk's "Powerwall" can also provide electricity for your home when your mainstream connection goes down.

Tesla Motors' Powerwall is not the first system to offer DC power for those who want to go "off the grid." Thousands of homes in remote locations have done this for decades with a combination of solar, wind, geothermal, and even biomass (turning manure into methane) energy. But the Powerwall system not only produces DC power; it also converts it to AC so that all your home appliances can operate. Each Powerwall (the initial units for sale) can store up to seven kilowatts, which is adequate for the average home. However, in order to gain wide acceptance, the system will have to be priced affordably, be adopted by homebuilders and remodelers, and not be blackballed by the still-powerful utility industry. That may take some time.

Musk's radical innovation quiver is far from empty. He has proposed developing a "Hyperloop" transportation system that would use magnetic levitation to propel passengers some 700 miles per hour in a closed tube from Southern to Northern California (and perhaps elsewhere). Unsurprisingly, it also bears some resemblance to an idea described in Tesla's autobiography:

> “Another one of my projects was to construct a ring around the equator which would, of course, float freely and could be arrested in its spinning motion by reactionary forces, thus enabling travel at a rate of about one thousand miles an hour, impracticable by rail.”

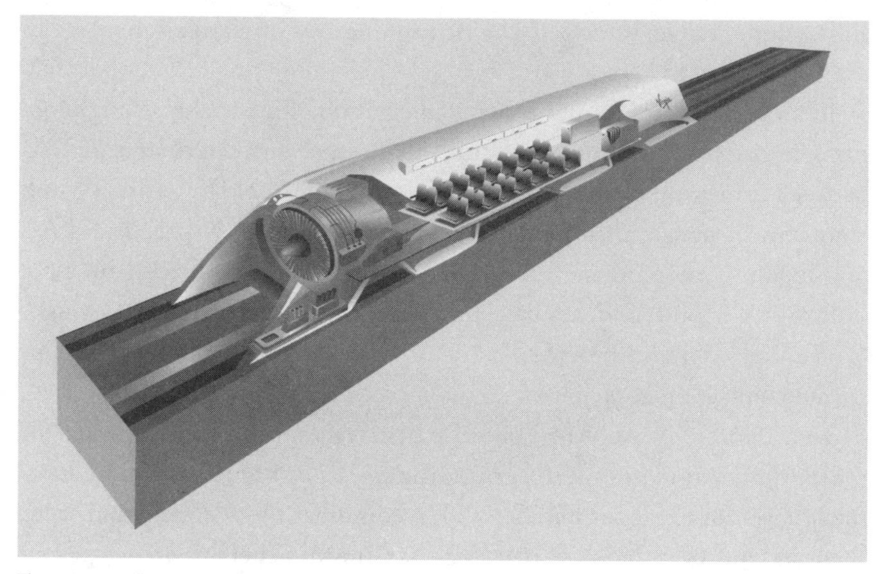

This concept illustration of Musk's Hyperloop, created by artist Camilo Sanchez, evokes one of the hypothetical technologies Tesla mentioned in his 1919 autobiography.

From Tesla to Google

The Tesla narrative, including his influence on Musk, has sent shock waves through the world of technology. For those embracing large ideas on a global scale, Tesla has become a patron saint—a Saint Sebastian, pierced with arrows of misfortune—who was martyred for his clash with capitalism.

Larry Page was twelve years old when he read John J. O'Neill's biography of Nikola Tesla. The future co-founder of Google was said to have cried when he finished the book, feeling a personal psychic pain that a man of such grand ideas could end his life in such a shabby way.

The son of Michigan State University computer science professors, Page was familiar with technology ideas and encouraged to explore the world. He tinkered, experimented, and played with computers. In 1995, after reaching Stanford University and its world-famous computer science doctorate program—the wellspring for all things technological—he met up with Russian-born fellow PhD student Sergey Brin. On September 4, 1998, they began Google, basing their operation at first out of the Stanford campus and then out of a garage. They quickly outgrew the garage and, after raising $1 million from friends and family, moved above a bike shop in nearby Palo Alto.

A loose, creativity-based management style guided the early days of Google, which has since been parceled off and put under a holding company called Alphabet. Employees were encouraged to have fun, to work on their own projects, and to not "be evil." Although they were up against the biggest players in technology—Microsoft and Apple—a sense of playfulness and joy infused the "Googlers" with a sense of purpose and abandon.

Buoyed by Tesla's motivation to create ideas that would benefit humanity, Page based his management philosophy on ethics. Considered a new breed

The Mountain View, California, corporate headquarters of Google Inc.—known as the "Googleplex"—is a unique, sprawling work campus with laundry rooms, swimming pools, eighteen cafeterias, and even a dinosaur skeleton.

of "regenerative entrepreneur," he adopted principles he had learned in a leadership program as an undergraduate at the University of Michigan. The motto of the college's LeaderShape Program is "Lead with Integrity. Disregard the Impossible. Do Something Extraordinary." Google employee Chade-Meng Ten, who holds the Google title of "Jolly Good Fellow," writes that the company has "a culture of compassionate concern for the greater good. Echoing Tesla's global spiritual goal of bringing peace to the world, Google—now Alphabet—employees are on a mission."

Through affiliate companies, Alphabet focuses on integrative systems to improve quality of life everywhere. On a microscale, its Nest thermostats sense when users need cooling and heating to save energy costs. On a larger scale, the company hopes to create Internet-linked "smart cities" that control resources efficiently through its Sidewalk Labs. Its Google Glass eyeglass/computer interface practically puts virtual reality and the power of the Internet right into your optic nerves. Other Alphabet ventures include self-driving cars and health-care and life-extension research.

With more than $70 billion in annual revenues and a stock-market value among the highest of any public company, Google/Alphabet may move even further into the meta-integration of personal technology, global systems, and energy management. It's not only creating its own clean energy for its massive computer servers; its larger mission is to defeat the planet-killing paradigm based on fossil fuels.

There's no question that Google/Alphabet owns Tesla as a spiritual godfather, but where Tesla failed as a businessman, Page and his companies are succeeding. The ideas that were planted like bamboo shoots a hundred years ago are becoming a dense forest of possibilities.

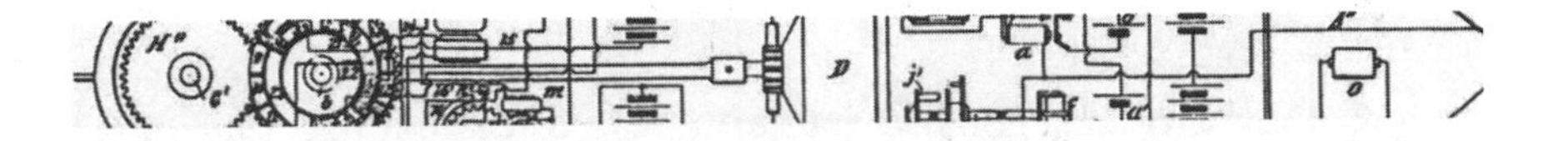

TESLA AND OTHER INNOVATORS: WHAT THEY HAD IN COMMON

Few people understand the spirit of innovation more than Walter Isaacson, who—as mentioned in chapter IX—has penned definitive biographies of Benjamin Franklin, Albert Einstein, and Steve Jobs. His insights on these creative geniuses culminated in his book *The Innovators: How a Group of Hackers, Geniuses and Geeks Created the Digital Revolution*. It's Isaacson who has synthesized the development of technology from the early days of the Enlightenment to our digital present.

Isaacson weaves a tale of unlikely innovators. Working his way through history to Jobs's Apple empire, Isaacson distills what made paradigm shifting creativity possible:

- **Creativity comes from the merger of the arts and sciences.** Leonardo da Vinci was a virtuoso draftsman and painter before he dabbled in his scientific pursuits. Jobs embraced arts and designs from around the world. Einstein played the violin. And Tesla used poetry and literature as inspiration for some of his most groundbreaking discoveries. Great innovators embrace the humanities and incorporate them into their spirit and work.

- **Creators have humility and tolerance.** On his deathbed, Einstein rued that he hadn't arrived at the "Theory of Everything." Benjamin Franklin welcomed everyone into his creative circle, which he called a "junto" that evolved into the American Philosophical Society. "If you're to be creative, you need to have a certain humility—you have to listen to people," Isaacson said when I heard him speak at the Chicago Humanities Festival in 2015.

- **Creators must be passionate and able to distort reality.** You have to be able to break boundaries and blow up existing paradigms. Why did Apple, once known as a second-tier computer company, get into the mobile phone or music downloading business? Certainly many skeptics told the Apple team they had no business in those industries, although history has shown how they came to dominate and disrupt the old models. Isaacson tells how Jobs pressed Corning Glass to develop its unique, scratchless "gorilla" glass for the iPhone. The Corning executives said they couldn't meet Jobs's seemingly unrealistic deadline. "Jobs sat and stared at them. They did it." Jobs and his team were able to distort reality to reach a goal in much the same way as Tesla envisioned a rotating magnetic field and built a system (and century) around it. Persistence combined with knowledge can be an unstoppable force.
- **Creation is a team sport.** According to Isaacson, the most potent advances come from an engaging group dynamic. All the great innovations of our tech age came from teams of people, not all of whom were geniuses. Collaborations of teams of dissimilar individuals created things like lightbulbs, energy systems, the Internet, cellphones, and smart cities. The individuals didn't always get along, but they managed to work together to achieve a common goal. Some of Tesla's most globally important work, including his alternating current, came from collaborations with other engineers. You need an array of talents, networked through the world of knowledge and experimentation, to come up with something useful that changes the world. The impossible becomes probable in the mind of a true creative soul, but you need to work with others to transform your passion into reality.

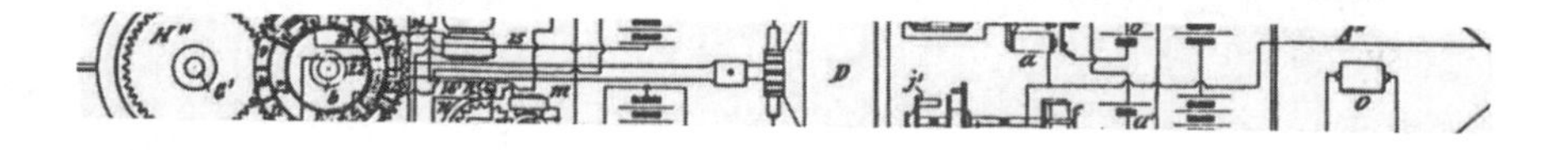

The Fourth Industrial Revolution

My last visit to the Hotel New Yorker, where I was staying a month before I completed my research, actually had nothing to do with this book. I was attending a journalism conference sponsored by the Society of American Business Editors and Writers, a journalism group I've relished being a member of for more than a decade. I also was on a panel talking about the evils of college loan debt (the subject of my first ebook, *The Debt-Free Degree)*. Without affordable education, we're headed for darkness, not a brighter future.

Little did I suspect that I would find yet another Tesla link at this event. Daniel Doctoroff, the deputy mayor of New York City under Michael Bloomberg and a former executive at Bloomberg LP, the financial information company, was speaking about the "connected city" as chief executive of Sidewalk Labs. In their pursuit of a system featuring public Internet access points throughout the city, Doctoroff and his team are exploring the "disruptive" possibilities of making cyber connections a widely shared resource that's available to anyone.

"Our ability to share [cyber] resources will create more prosperous cities," Doctoroff predicted, noting that most of the world's population now lives in urban areas. "The integration of data and technology will make cities more livable." Offering free wireless Internet access throughout the city will seed this transformation, Doctoroff asserts. That would augment the larger vision of creating "smart" cities that use big data for more efficient use of energy, space, and resources.

Where are energy and information needed, and how can they be better directed? How can cities better obtain energy when they need it most—such as during megastorms like Hurricane Sandy, which flooded and darkened nearly half of Manhattan in 2012? In a time when climate change profoundly threatens our existing infrastructure, the wireless sharing of information, power, and resources is essential. Such is the mission of Sidewalk Labs, funded by Alphabet Inc. and blessed by Tesla-inspired Larry Page.

A systems approach to solving widespread urban resource issues will unite the power of supercomputing, energy networks, the Internet, and big data into a meta-integration that will make urban life more feasible and sustainable. Greener cities, transportation, vertical farms, robotics, and other vital systems are emerging from the meta-integration of Tesla's worldview.

But it may take some interesting turns away from the "big systems" thinking of Tesla and Insull's central power station–driven grid and become more evenly distributed. As the world has crossed the economic threshold of affordable clean energy (as compared with fossil fuels), building owners will be creating and managing their own power, creating *microgrids*. For example, a few blocks away from where Insull fired up the world's first large-scale turbogenerator, the Illinois Institute of Technology in Chicago has its own microgrid, powered by wind and solar energy. The college's Robert Galvin Center for Electricity Innovation is leading research in this "greener and cleaner" direction.

Potentially all homes and buildings will be able to provide their own power for "prosumers" (thank you, Alvin Toffler) of the future, with surplus electricity feeding large batteries and electric vehicles. If this green energy system scales up, cities may be able to reduce their reliance upon fossil fuels on an exponential scale.

Independent, virtually wireless power sources will change the way homes, transportation systems, and entire cities are built—and reconfigured—to what the World Economic Forum is calling the "Fourth Industrial Revolution." This paradigm-changing wave is melding artificial intelligence, cybernetics, the Internet, and genomics in a massive chimeric merger. For example, houses can automatically "think" to adjust to energy needs, weather changes, and possibly even climate change. High-end "smart" homes do this already—homeowners can program them from afar or shut them down for vacations.

Embedded sensors in your body will tell you when you're getting sick and connect with independent computers to tell you how to heal faster. You may be even able to reprogram your genes with the help of an online-enabled supercomputer to help you avoid cancer, diabetes, or heart disease. Everyone who's connected in this meta-integrated world would live longer and happier—unless, of course, a giant intelligent network emerges after a "singularity" (see Ray Kurzweil's book *The Singularity Is*

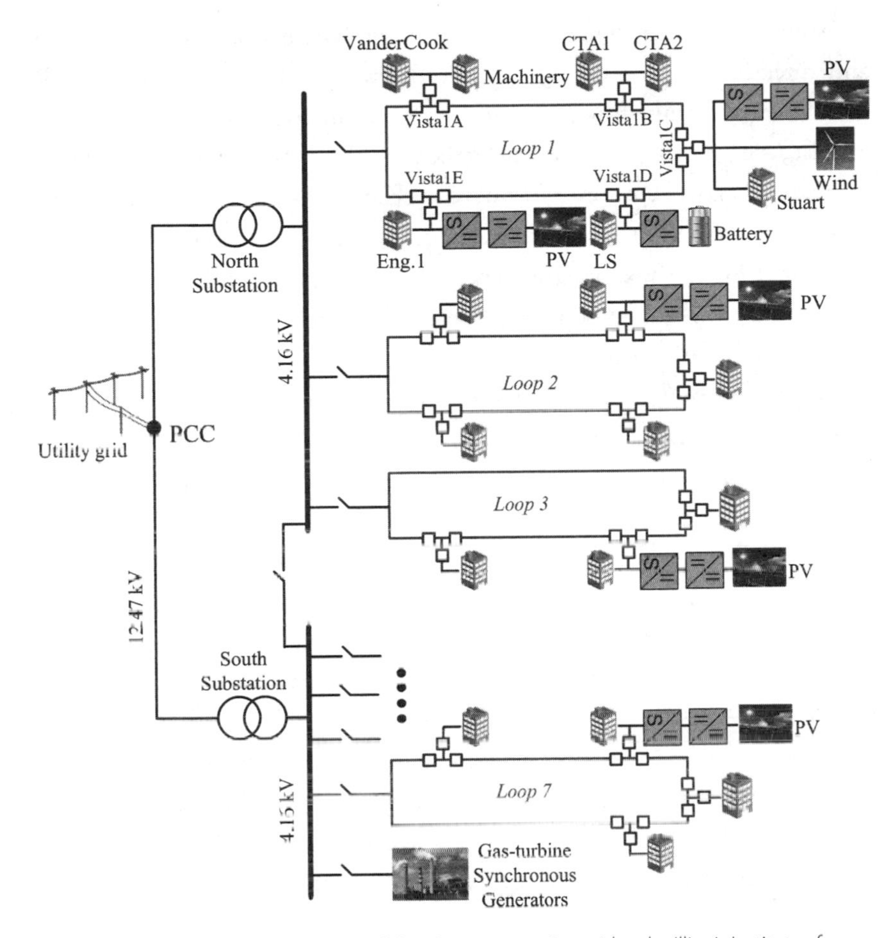

This diagram shows the electrical loops of the clean power microgrid at the Illinois Institute of Technology in Chicago.

Near) and decides that the planet is better off without us. The prospect scares even Musk and other tech titans. Having become the face for land and space systems innovation, Musk has promised that wireless electricity will become a reality. Like many technologists, he also advocates an ethical framework for implementing artificial intelligence and courageously offers to become a guardian of humanity when artificial intelligence becomes "an existential threat."

Why can't an intelligent network provide wireless power, information, and directions to homes and vehicles, anyway? Experiments in the U.K. and Korea are embedding electrical emitters in roadbeds to power wireless electric vehicles. It's not far-fetched to think that the transportation infrastructure of the future won't have overhead power lines and won't need fossil fuels. Also gone would be diesel engines in buses, trains, and trucks. Of course, for all these systems to help the planet, the power would have to come from clean energy sources. Wireless induction motors—Tesla's brainchild—now, *that's* the ticket!

What's ever more important to our civilization now isn't another weapon of mass destruction. We need some technological breakthroughs to stem climate change, which will impact all of us and trigger even more global conflicts and devastating wars over resources.

Tesla was so far ahead of his time that the energy that he unleashed in the form of ideas is still with us and transforming our relationship with the earth. A lot of his ideas seem far-fetched—and they especially seemed so when he formulated them more than a century ago—but geothermal power *is* now a reality. We *may* be able to transmit power through the earth. Wireless power *could* become widespread globally, especially if we're able to transmit power from solar panels floating in space. Electricity and magnetic fields *are* used for healing and diagnostic purposes.

Telegeodynamics may also come to fruition someday. Maybe Tesla heirs like engineer Eric Dollard, working in the Mojave Desert without phones or Internet connections, will someday use Tesla's technologies to predict earthquakes. I'm in no position to evaluate the feasibility of any of these technologies, but if there's an overarching lesson to Tesla's life and legacy, it is to keep an open mind.

If we want to find ways to save humanity from extinction, we have to leverage the infinite possibilities of meta-integration and artificial intelligence by coming together from different spheres. Technologists need to be talking with humanities scholars. Designers need to be talking with linguists. We need to spend less time on social media and more time interacting with people in face-to-face social situations. "The thing that makes [humans] special is that we're mind-readers," notes Nicholas Epley, a neuroscientist

This rendering depicts the plans for the development of a "smart city" in the Nansha district of China.

at the University of Chicago. "It's social intellect that makes us unique." In his book *Mindwise*, Epley writes that "our minds are indeed capable of extraordinary feats, but it is false that 90 percent is untapped potential. These unconscious processes are tapped every second of your life, governing nearly all of your behavior in a reasonably adaptive fashion." Tesla's spirit has permeated our consciousness in many ways, and his vision seeps into ongoing projects to improve civilization every day.

Although more sustainable systems are clearly essential to our survival, we need a restored faith in ourselves and the idea that we can invent what we need and change our behavior to make this a better world. Turning the conventional wisdom upside down and reinventing cultural attitudes is one step. Ethics, self-reliance, and faith in ourselves have to be part of the equation. We also have to renounce violence in all forms and find better ways of resolving conflict, greed, and soul sickness without weapons.

The most enduring innovators never give up on the uniquely human ability to redeem and reenvision the world. Tesla, to his everlasting credit, tied innovation and community building to world peace. In a short, powerful section of his autobiography titled "The Road to Permanent Peace," written more than a century ago, he says:

> “What we now want most is closer contact and better understanding between individuals and communities all over the earth and the elimination of that fanatic devotion to exalted ideas of national egoism and pride, which is always prone to plunge the world into primeval barbarism and strife.”

One Last Visit to the Hotel New Yorker

Just before I finished the notes for this book, I ventured back to the Hotel New Yorker for one last peek before I headed home to Chicago to begin writing. I wanted to check out Room 3327 and take a digital snapshot of the Tesla memorial plaque on the door. As I was adjusting my smartphone camera, the door suddenly swung open, startling me. Cartoonist Teresa Roberts Logan (aka "The Laughing Redhead," an apt nickname) stood in the doorway.

At first, I was embarrassed that Logan and the two other people in the room—illustrator Carolyn Belefski and filmmaker/photographer Joe Carabeo—might think I was some sort of creepy stalker. So I explained quickly that I had no intention of knocking on the door and that I was writing a book on Tesla. Then I asked if they knew who Tesla was.

They all smiled.

Not only did they know about Tesla, but they were definitely "feeling Tesla's energy" in the room. Since they were in town for the ComicCon—an annual celebration of the human imagination—I didn't have to tell them Tesla's story. With their creativity and adventurous spirit, they were already sharing his power with the universe.

 10 Be Chimeric

Change is inevitable and difficult to navigate. For positive change to enter into our lives, we must be chimeric—that is, we must incorporate and embody the different parts of our intelligence and personality that we haven't yet embraced. And there's so much to recognize and develop! It's no longer simply a "left brain" vs. "right brain" world. Harvard professor Howard Gardner, a global expert on intelligence and education, recognizes multiple forms of intelligence: verbal-linguistic (communication, learning new languages); math/logic (solving complex problems, critical thinking); musical (rhythmic and harmonic); visual-spatial navigation (dancers, artists, and pilots have this); bodily-kinesthetic (athleticism); interpersonal (working and interacting with others); intrapersonal (introspective); naturalistic (understanding how nature operates); and teaching. We all have different levels of each type of intelligence, but to make the most of our potential as human beings in a changing world, we need to integrate some of each form.

What does this mean in practical terms? Take on what you don't know. Learn something new. Understand how the body channels energy. When I saw tai chi master Chungliang Al Huang lecture a year before I finished this book, he helped me to understand how tai chi (and its compressed form of energy exercise, chi gong, which I practice every day) could amplify the mind, body, and spirit. "Open the gate to see the mountain," Huang tells you while doing exquisite Chinese calligraphy and dancing. Where does this energy come from? Well, from the universe, as Tesla understood. How do you enhance it? By laughing, dancing, doing math games, teaching, making music, and exploring sides of yourself that you don't really know. It's then that you discover the fertility of relationships and who your friends are.

Tesla is a shining example of the creative chimera. He integrated linguistic intelligence with his command of several languages, intrapersonal intelligence with his ceaseless introspection, visual-spatial intelligence with his visualization practice, and naturalistic intelligence via observation and excursions in the mountains of Europe and America. He built upon his natural mathematical-logical intelligence with efforts to improve his bodily-kinesthetic, musical, teaching, and interpersonal intelligence as he swam in the Seine, socialized with great artists and scientists, entranced crowds with his AC current and remote-control boat, and expounded on his colorful life and myriad ideas in personal letters and respected magazines.

In the future, a physical chimeric integration will more intimately merge us with machine intelligence, for better or worse. For some, it's already happened. Chips inside our bodies may be able to reprogram genes or short-circuit heart disease or diabetes. Already we have artificial limbs and joints, some of which are connected to our neural circuitry. I'm not sure when—or if—we'll hit Ray Kurzweil's "singularity" and become the slaves of machines. Will a machine ever be able to compose a Shakespeare play, Bach fugue, or T.S. Eliot poem? Perhaps computers can copy such things, but they still have the "divine" spark of their makers. That's where you come in.

Feel the energy all around you and within you. Use it to broadcast your unique combination of talents and contributions to the world. Wander around in an unknown place. If you don't do any art, try doodling. If you can't run, walk. Balance your checkbook. Cook something exotic. Teach something you know. Try to understand what's going on with the earth. Think of yourself as the sum of many beautiful parts. Disrupt the universe with your luminosity.

EPILOGUE

A HAUNTING AIR

On an island in Lough Derg, a lake in Donegal, Ireland, there is a basilica and monastery. People have been coming here for more than 1,000 years to do pilgrimages in their bare feet.

I have seen these acolytes first hand, climbing holy, stony mountains like Croagh Patrick further down the Irish coast, their feet sore and bloody, attempting to purge their sins the way St. Patrick was said to have done. My wife and I have climbed this mountain as well (with shoes on, thankfully).

There is a song called "The Cross of Lough Derg." I first heard it played by the astounding fiddler Liz Carroll (who also composed the piece) at a concert at the Art Institute of Chicago. It is a simple melody of painful longing and regret.

As a violinist and singer in an Irish/Roots band (The Prairie Fogg), I had never heard the song before, yet some subterranean emotional current caused me to sob. Maybe it triggered some unacknowledged regrets that ran as deep as a cave river. Or maybe it lay in Liz Carroll's passionate rendition. Months later, upon hearing the song again in a recording, it made me think of Tesla.

It could be that Tesla, alone with his pigeons and ignored ideas, felt the kind of longing that exiles feel after being separated from their native land for fifty years, the kind of subdued passion that would trouble them in their last days. When he died, his country had been overrun by *Ustase*, fascist Croatian Nazis who furiously murdered thousands of Tesla's fellow Serbs, along with Roma and Jews.

This photograph, taken by Andreas F. Borchert with kind permission of the prior Monsignor Richard Mohan, shows the basilica designed by William Scott (1871–1921) and built on Station Island in Lough Derg to mark the pilgrimage site of St. Patrick's Purgatory.

Would Tesla have known of this slaughter? Since he had a connection to his country through young King Peter—who later settled in the United States in exile—perhaps they spoke of the genocide. At the time, Tesla's life circuit was open: His inventions were running the world's electrical grid, and his unifying universal system was lying somewhere in his extensive notes. His pious mother and father were long gone. His sweet and beloved brother Dane was still haunting his consciousness. More visions, less sleep.

What had he done? Wired a world that was so disconnected and fragmented that wars were being waged across great continents and vast oceans. Although he had a way to send power through the earth and heavens, his theories could do nothing to short-circuit the lightning that was setting nearly all of civilization on fire.

And so the regret. There was no unifying theory that would bring peace before he died. The sad melody played on as the world did its penance after the horrors. Yet Tesla's ideas remained as verdant islands of hope, tucked away in sacred lagoons where only saints endured the temptation of the devil.

How do we transcend the evil complacency of the world to tackle some of the globe's greatest problems? We certainly can apply creativity and generosity of spirit—as Tesla did—but we also need to grapple with some of the Faustian bargains involved. The Fourth Industrial Revolution, which is the meta-integration of the Internet, artificial intelligence/robotics, big data, and genomics, is the ultimate chimeric transmutation. We are merging ourselves into the *uber*-machine age, yet we need to acknowledge some of the ethical consequences.

Physicist Stephen Hawking, for one, warns that such a transformation "creates new ways that things can go wrong." He cautions, for example, that we may be genetically engineering superviruses, and we are not fully aware of the ongoing threats of nuclear war and global warming. The planet's health should be our major concern—*not* whether a smartphone can download videos quickly or we can engineer better babies. Pope Francis warned in his 2015 encyclical *Laudato si'* that climate change should be seen as a global spiritual crisis as well as an ecological one. People's homes will disappear and millions will die during ongoing extreme climate events. We need to adopt what the pope calls an "integral ecology" that combines science, technology, ethics, and compassion for the world's most vulnerable populations.

How do you offer compassion to the most vulnerable while addressing the burgeoning promise, delight, and *threat* of technology?

It will be hard to create an ethical framework when so much is happening so quickly. More than 24,000 robots, for instance, were delivered to customers in 2014. An estimated one million (mostly *toy*) drones were given as Christmas gifts in December 2015. Millions of robots are now being designed to help the elderly and disabled. Robotic, driverless cars are being developed by Google, Tesla Motors, and many other manufacturers.

All told, there are more than one million "working robots" on the job today, according to the World Economic Forum (WEF). Meanwhile, drones are being used to monitor and assassinate enemies of the United States all over the world, in addition to performing other, less lethal tasks such as monitoring the effects of global warming. Although the total number of military drones is unknown, there are more than 30,000 companies that contract with the Pentagon to supply them. And they are getting more ubiquitous and deadlier by the day. There are cameras and programs everywhere, watching our every move.

Although Tesla clearly conceived of weapons through his robotics research and "World System" development, he promoted the idea of *defending* countries with his technology. I think he would have endorsed a "Department of Peace" for this purpose.

Another threat we are facing is the displacement of human energy and employment as the meta-integration of technology gains steam. One shocking prediction by the WEF is that up to *half* of all jobs "are likely to become computerized in the next 10-20 years." The WEF, in what seems like a conservative estimate, projects that five million jobs in developed countries will be lost over the next half-decade through widespread automation. While that's good news for robotics/AI designers and manufacturers, it's bad news for the rest of us.

Ultimately, the ongoing meta-integration may exacerbate the growing economic inequality across the globe. As Robert Shiller, a Yale economics professor and Nobel Prize winner noted at the Davos Forum in 2016:

> You cannot wait until a house burns down to buy fire insurance on it. We cannot wait until there are massive dislocations in our society to prepare for the Fourth Industrial Revolution.

Can mammoth breakthroughs like quantum computers, universal wireless power, robotics, and massively shared information help create entirely new systems to avert catastrophic climate change while adding decent, new jobs? There's some reason for hope. Buildings and homes can now monitor themselves through cybernetic systems to lower their energy costs and carbon footprints. So-called "zero energy" homes can create their own power and use the earth's thermal resources for on-site heating and cooling. An MIT-trained engineer and inventor I know, Victor Zaderej, is not only designing such homes, but also perfecting low-energy lighting systems that dramatically lower lighting costs for any home or other building. These systems can even be used for indoor growing systems and are bringing light to some of the darkest places on the planet.

At the core of Zaderej's creative core is "curiosity, a willingness to make mistakes, a basic understanding of how things work and the ability to build knowledge over the years." He's a modern-day Tesla among hundreds of thousands of makers, tinkerers, hackers, and coders who are experimenting every day.

On a broader scale, though, the world has crossed a golden, promising threshold, which is even better news. Alternative energy such as solar and wind power is now economically competitive with fossil fuels. Exoskeletons help the disabled function again, and gene editing promises to "edit out" biological codes that make us sick. Artificial intelligence isn't a pipe dream; it's there every time you log on, for better or worse. Rockets are reusable and becoming cheaper. Tesla would have cheered this progress, too.

"Salvation from climate catastrophe," opines economist and columnist Paul Krugman, "is something we can realistically hope to see happen, with no political miracle necessary."

How do we integrate technological progress with our humanity and spiritual needs? In many ways, the ascetic and celibate Tesla was the essence of a spiritual scientist, seeking peace and enlightenment and sacrificing his entire being to discovering the truth in lightning, a Prometheus of the redeeming and often destructive nature of energy itself.

But how do we redeem and enhance our humanity when it becomes so easy to consume through digital payments and online robots that find more things for us to buy, further ravaging the planet? How can we use the disruptive promise of technology to improve our *quality* of life and not the *quantity* of our materialistic lusts? How do we maintain our privacy in a world in which our every transaction is monitored?

In reclaiming a spiritual ecology and economics—one that broadly connects us akin to Tesla's "World System"—we can start talking with each other to bridge the ethical and human capital divide that technology has created.

"With digitalization," observes futurist author Jeremy Rifkin, "we have the tool to connect the human race in increasingly inclusive networks so that we can begin to think and behave in an extended human family for the first time in history."

Collective compassion and a spiritual economics must be the guiding ideas behind technology's many offerings. We need to listen carefully and act upon global concerns using the meta-communication networks we've created. The global community needs to find a consensus on shared problems and not pander to individual desires. And humanity must heed what Sigmund Freud observed in the 1920s when he wrote in *Civilization and Its Discontents*:

> Human life in common is only made possible when a majority comes together which is stronger than any separate individual and which remains united against all separate individuals. The power of this community is then set up as "right" in opposition to the power of the individual, which is condemned as "brute force."

More than ever, we need to produce what Tesla called "terrestrial motions at a distance" to disrupt the disturbing proliferation of violence (from American cities to the Middle East), economic inequality, and global resource depletion. But this techno-cultural shock wave needn't create actual earthquakes. We can employ our meta-integrated world systems to chimerically and positively transform our thinking and social systems. That's a powerful idea that has infinite force.

REFERENCES & ABBREVIATIONS

I researched several archives in preparing this book. Many sources contained pieces of other collections, so documents I reviewed from one archive may have been acquired from another. I've also standardized some name spellings, especially those from Eastern Europe, which have multiple variations. Here are the major sources of Tesla material that I used in this book:

(FBI) Archives of the Federal Bureau of Investigation, Washington, D.C.

https://vault.fbi.gov/nikola-tesla

Although lacking information on the final location of Tesla's papers, this collection, beginning with memos from the day of his death in 1943, runs for several decades and is the principal source for the government's tracking of Tesla's records, which were confiscated upon his death. The file was obtained through a Freedom of Information Act (FOIA) request, but some of the key memos are heavily redacted. This is the only collection I encountered that was missing key elements.

(GL) Paul Galvin Library, Illinois Institute of Technology, Chicago.

http://library.iit.edu/search?keys=World%27s+Columbian+Exposition

This was my main resource for materials and images from the 1893 World's Columbian Exposition.

(HHC) Sen. John Heinz History Center, Pittsburgh.
http://www.heinzhistorycenter.org/collections/westinghouse

Keeper of the Westinghouse collection, this generous archive also has papers from Tesla archivist Leland Anderson, Columbia University, and copies of documents from the Tesla Museum.

(MI) *My Inventions: With the Correspondence between Nikola Tesla and Hugo Gernsback* by Nikola Tesla, edited by Hugo Gernsback and Vladimir Jelenkovic (Belgrade: Tesla Museum, 2006).

This is often called the "autobiography" of Tesla, although it's a collection of articles and letters from 1919. In his own words, Tesla recounts his life and inventions up to 1919. Though incomplete, it's a widely used account of Tesla's early days from birth to the creation of his "magnifying transmitter."

(ML) The Morgan Library & Museum, New York City.

http://www.themorgan.org/collection/archives

Housing the papers of J.P. Morgan and his son, among other documents, the Morgan Library was the principal source for the Tesla-Morgan letter series.

(SIP) Samuel Insull Papers, Cudahy Library, Loyola University Chicago.

http://www.luc.edu/media/lucedu/archives/pdfs/insull1.pdf

Loyola contains most of Insull's papers. I also acquired some Edison-Insull documents through the Edison papers, although I couldn't find a single document relating to Tesla in this vast cache of digitized Edisonia. See http://edison.rutgers.edu/.

(TFC) Twenty-First Century Books.

http://www.tfcbooks.com/tesla/contents.htm

This is a prime online source for Tesla's writings and newspaper pieces about him from 1888 to 1944.

(TM) Nikola Tesla Museum, Belgrade, Serbia.

http://www.tesla-museum.org/meni_en/nt.php?link=arhiva/a&opc=sub5

The world's best collection of documents, letters, and artifacts from the inventor. I viewed document copies through restricted, remote computer access. The museum also provided a complete list of Tesla's patents and an index to most of its documents.

(TMP) Tesla Memory Project: "The Essential Nikola Tesla: Peacebuilding Endeavor."
http://fim.rs/en/essential-nikola-tesla-peacebuilding-endeavor/

These brief observations, from which I've quoted in the epitaphs of several chapters, were prepared by the UNESCO Center for Peace and the Energy Innovation Center TESLIANUM, under the direction of Aleksandar Protic.

(TU) Tesla Universe.
https://teslauniverse.com/nikola-tesla/timeline/1856-birth-nikola-tesla#goto-320

During my research for this book, I made frequent use of this omnibus website on Tesla materials—particularly its well-organized timeline of major events in Tesla's life and beyond.

Books about Tesla

(CWF) ***Chicago's 1893 World's Fair*** by Joseph DiCola and David Stone (Charleston, S.C.: Arcadia Publishing, 2012).

A compact guide to the images and events at the fair.

(IRW) ***The Inventions, Researches, and Writings of Nikola Tesla*** (New York: Fall River Press, 2012).

This is a fairly comprehensive collection of Tesla's technical papers with a short biography and a section on his exhibit at the 1893 World's Fair.

(MOP) ***The Merchant of Power: Sam Insull, Thomas Edison and the Creation of the Modern Metropolis*** by John F. Wasik (New York: Palgrave Macmillan, 2006).

This exploration of the life and work of Samuel Insull led me to Tesla and is the main source of my knowledge on the Insull-Tesla relationship.

(PG) *Prodigal Genius: The Life of Nikola Tesla* by John J. O'Neill (Kempton, Ill.: Adventures Unlimited, 2008/3rd edition).

The first biography of Tesla, by a journalist who knew him.

(TI) *Tesla: Inventor of the Electrical Age* by W. Bernard Carlson (Princeton: Princeton University Press, 2013).

In this most recent of the major Tesla biographies, Carlson presents Tesla's work in the context of the history of science and technology.

(TML) *Tesla: Master of Lightning* by Margaret Cheney and Robert Uth (New York: Metrobooks, 1999).

This is the companion book to the excellent PBS series of the same name (see www.pbs.org/tesla/).

(TMOT) *Tesla: Man Out of Time* by Margaret Cheney (New York: Touchstone, 1981).

Although dated, this is a fine introduction to Tesla that introduced some new research at the time.

(W) *Wizard: The Life and Times of Nikola Tesla: Biography of a Genius* by Marc Seifer (New York: Kensington Publishing, 1998).

This is one of the most detailed biographies of Tesla.

(WE) *Tesla: The Wizard of Electricity* by David J. Kent (New York: Fall River Press, 2013).

A lavishly illustrated primer on all things Tesla.

BIBLIOGRAPHIC ESSAY

Please note that I relied extensively upon the biographies of Carlson (TI), Cheney (TMOT), O'Neill (PG), and Seifer (W) for the core narrative. Many of Tesla's own reflections throughout the book come from *My Inventions* (MI), originally a 1919 magazine article in *Electrical Experimenter* by editor Hugo Gernsback (after whom the "Hugo" science fiction award is named).

Introduction

I discovered the original letter that triggered my research for this book while I was researching *The Merchant of Power* (MOP), my biography of Samuel Insull. The letter from the Insull archives (SIP), which contained a short summary of Tesla's "Art of Telegeodynamics," was dated March 18, 1935, and sent to Insull from Tesla at the Hotel New Yorker. I didn't find Insull's reply to the letter. Insull died in 1938.

Also see "Nikola Tesla Revolutionized World with Grid, Wireless" by Scott Smith, *Investor's Business Daily*, Dec. 7, 2015.

The chimera appears in the Greek myth "Bellerophon and the Flying Horse," *Usborne Greek Myths* (London: Usborne, 1999), pp. 54-57. It's a great allegory for embracing the different aspects of human personality and adapting to change.

Chapter I

Milena Bajich, who has a doctorate in clinical psychology, has been studying Tesla for decades. I interviewed her on Sept. 4, 2015, in Chicago. I cite research from her presentation "Nikola Tesla: The Man Who Could Not Stop Inventing" given at the Tesla Conference in Philadelphia in 2013. I also clarified some of her main points through subsequent emails.

Tesla's commentary on his eccentricities and the value of introspection are from his autobiography (MI). Unless otherwise cited, Tesla's own reflections, recollections, and observations are from this book.

My account of the FBI's involvement in Tesla's last day is largely derived from the memos in the Tesla FBI file (FBI), originally obtained through the Freedom of Information Act. I also consulted the obituary in the *New York Times*, dated Jan. 8, 1943 (http://ethw.org/images/4/40/Tesla_-_obituaries_for_tesla.pdf).

A prime source used throughout the book on Tesla's missing papers is from Marc Seifer, particularly his article "Nikola Tesla: The History of Lasers and Particle Beam Weapons" in *Proceedings of the 1988 International Tesla Symposium* and his superb biography *Wizard* (W). Dr. Seifer also responded with clarifications to several emails I sent in the last two months of writing this book.

For images and explanations of Tesla's last efforts, Cheney and Uth's *Tesla: Master of Lightning* (TML) was referenced. The book was a companion to a PBS series of the same name. The network's website has a section for the series that contains excellent resources, including an article on Tesla's missing papers (http://www.pbs.org/tesla/ll/ll_mispapers.html).

Multiple visits to the Hotel New Yorker over the course of three years were the source of my information about Tesla's last years. Also see TU.

Tesla's "death ray" and "telegeodynamics" concepts were publicized in several New York papers from the middle to late 1930s. Twenty-First Century Books (TFC) has reproduced most of these articles online. The article describing the particle beam weapon, "Soviets Push for Beam Weapon," appeared in *Aviation Week and Space Technology*, May 2, 1977 (http://www.larouchepub.com/eiw/public/1977/eirv04n19-19770510/eirv04n19-19770510_040-aviation_week_magazine_soviets_p.pdf).

Tesla's last days were recorded by journalist John O'Neill in *Prodigal Genius* (PG), the first biography of Tesla by a man who reported on him.

Chapter II

The epigraph is from the TMP.

I saw sections of Leonardo's notebooks in the Barcelona Maritime Museum in August 2007. I also referenced Fritjof Capra's excellent *Learning From Leonardo: Decoding the Notebooks of a Genius* (San Francisco: BK Books, 2013), p. 25; *DaVinci's Ghost* (New York: Free Press, 2012) by Toby Lester; *Leonardo DaVinci* (Cobham, UK: TAJ Books, 2004); and Harold Bloom's priceless *Shakespeare: The Invention of the Human* (New York: Penguin, 1998), p. 666.

Tesla's autobiography (MI) is the source of his commentary on intellectual curiosity and powers of observation, along with the influence of Kelvin on his work.

Franklin's quote from his electrical experiments is found in the collection of letters titled *Experiments and observations made at Philadelphia in America by Mr. Benjamin Franklin* (Ecco Press, 1751). Another indispensable guide to Franklin is Walter Isaacson's *Benjamin Franklin: An American Life* (New York: Simon & Schuster, 2003).

The background on Faraday is from volume 45 of the Great Books series, published by the University of Chicago, 1952.

The origin story of *Frankenstein* is described in "The Poet, the Physician and the Birth of the Modern Vampire" by Andrew McConnell Stott (http://publicdomainreview.org/2014/10/16/the-poet-the-physician-and-the-birth-of-the-modern-vampire/).

Chapter III

My poem is titled "The Rising Sickness."

Tesla's earliest experiences are recounted in his own words in *My Inventions* (MI). These include numerous reflections on his visual "affliction" and its transformation into an intellectual asset.

Tesla's quotation describing a "mind journey" on page 53 was cited in Carlson (TI), p. 245. I also relied upon the biographies of Cheney (TMOT) and Seifer (W).

Milena Bajich's commentary comes from our Sept. 4, 2015, interview in Chicago. I cite research from her presentation "Nikola Tesla: The Man Who Could Not Stop Inventing" at the Tesla Conference in Philadelphia in 2013 and clarified some of her main points through subsequent emails.

A cogent introduction to the concept behind "flow" is John Geirland's *Wired* magazine piece "Go With the Flow," Sept 1, 1996 (http://www.wired.com/1996/09/czik/). *Flow: The Psychology of Optimal Experience* (New York: Harper Perennial, 1990), by Mihaly Csikszentmihalyi, is one of the seminal books on the study of creativity. I referenced pp. 74–77; pp. 208–213; pp. 214–221.

Another useful, though much more academic work that refers to Tesla's sensory abilities is *The Neuroscience of Creativity* (Cambridge, Mass.: MIT Press, 2013), edited by Oshin Vartanian, Adam Bristol, and James Kaufman (p. 180).

I derive my Faust quotation from the 1967 Modern Library edition of Goethe's *Faust*, based on an 1870 edition (p. 38). The quotation from Carlson analyzing the Faust passage comes from his biography of Tesla (TI), pp. 53–54.

For an accessible, almost nontechnical explanation of the workings of Tesla's motor and magnetic field, I was aided by "How does an induction motor work" online at http://www.learnengineering.org/2013/08/three-phase-induction-motor-working-squirrel-cage.html.

Chapter IV

The Robert Curl quote is from the Tesla Memory Project (TMP).

Tesla's first-person accounts of his travels as a young man, intense drive for success, and collaboration with Westinghouse come from *My Inventions* (MI), though many biographers suggest he may have embellished a bit. I put more stock in Seifer's (W) and Carlson's (TI) takes on his earliest years.

For the Edison/Pearl Street section, I relied upon Jill Jonnes's excellent *Empires of Light: Edison, Tesla, Westinghouse and the Race to Electrify the World* (New York: Random House, 2003); *Fleet Fire: Thomas Edison and the*

Pioneers of the Electric Revolution (New York: Arcade, 2003), by L.J. Davis; and Neil Baldwin's fine biography *Edison: Inventing the Century* (New York: Hyperion, 1995).

I also derived another perspective on Edison through the eyes of Samuel Insull from my *Merchant of Power* (MOP) and *The Memoirs of Samuel Insull: An Autobiography* (Polo, Ill.: Transportation Trails, 1992). Also instrumental in my understanding of Insull and Edison was Forrest McDonald's *Insull* (Chicago: University of Chicago Press, 1962).

Although there's only a plaque in lower Manhattan marking where the Pearl Street plant was located, some background is found in "Pearl Street Station" (http://ethw.org/Pearl_Street_Station). The *New York Times* revisited its own coverage of Edison and the development of incandescent lighting in "Describing Electric Light, Even Before Edison," Jan. 12, 2016.

The best accounts of the 1880s were in Cheney, Seifer, and Carlson.

For more on specific technical papers and his famous lectures, see Tesla's *Inventions, Researches, and Writings* (IRW).

The AC power plant in the Rockies was described in "Why Solving Climate Change Will Be Like Mobilizing for War," by Venkatesh Rao, *The Atlantic*, Oct. 15, 2015. Although the piece is about climate change, the author compares the technological quest to tackle global warming with the "Battle of Currents" in the 1890s.

The Kent quote is from *Tesla: The Wizard of Electricity* (WE), p. 201.

The note about the missing Edison Medal was in a letter dated June 25, 1955, from Kenneth Swezey, a friend of Tesla's during his last twenty years, to none other than J. Edgar Hoover. More than ten years after Tesla died, inquiries started coming into the FBI trying to locate the full contents of Tesla's room when he died in 1943. Swezey had been in the room the day of Tesla's death.

For the TeslAction I consulted Malcolm Gladwell's stellar book *Outliers: The Story of Success* (New York: Little, Brown and Co., 2008).

Chapter V

My poem to lead out this chapter is entitled "Bulging Lines of Force."

Some compelling descriptions of Tesla and Westinghouse's involvement in the fair and the Battle of Currents are in *Fleet Fire* and *Empires of Light*. For an insightful look into the atmosphere of the fair and Chicago, see Erik Larson's engaging *The Devil in the White City* (New York: Random House, 2003), although he mentions Tesla only once.

The lion's share of documents and images from the fair were found in the Paul Galvin Library at the Illinois Institute of Technology (GL), where I found extensive materials from contemporary sources, such as *The Book of the Fair* (1893) by Hubert Howe Bancroft. Also, DiCola and Stone's *Chicago's 1893 World's Fair* (CWF) was a good background source.

For insight into Westinghouse and his financial troubles, I referred to Quentin Skrabec's *George Westinghouse: Gentle Genius* (New York: Algora, 2007) and the Westinghouse archives at the Heinz Center (HHC), the source of all the Tesla-Westinghouse correspondence and the handful of letters from Twain to Tesla. Robert Underwood Johnson's memoir *Remembered Yesterdays* (New York: Little, Brown, 1923) also recounted his experiences with Tesla (and Twain) during that time. *Fleet Fire* and *Empires of Light* were also referenced for the travails of Westinghouse during that period. I get into more detail about the Morgan-GE consolidation in MOP. McDonald also covers it well in his *Insull*.

Another description of the fair, Henry Adams's reaction, and the H.G. Wells quote were in Jane Brox's *Brilliant: The Evolution of Artificial Light* (New York: Houghton Mifflin, 2010), pp. 150–152.

The image of Samuel Clemens, better known by his pen name Mark Twain, with the lightbulb on page 111 was shot March 3, 1894, at the height of Clemens's financial troubles. You can get a picture of Twain's frustration with his money-losing investments in various inventions in *Mark Twain Himself* by Milton Meltzer (New York: Bonanza, 1960) and Ron Power's *Mark Twain: A Life* (New York: Simon & Schuster, 2005).

Although there's some question on whether Tesla met Swami Vivekananda at the fair, he was influenced by the guru, who spoke at the fair and throughout the United States during that time. See http://www.teslasociety.com/tesla_and_swami.htm from the Tesla Memorial Society of New York and http://vivekananda.org/. Much study has been devoted to the extent to which Hindu literature has contributed to our knowledge of energy in the universe and our own bodies (chakras).

Tesla's quotes are from his autobiography (MI), in which he recalls his childhood vision of Niagara and love of Twain's novels, but his "the end has come" pronouncement comes from a *World Sunday Telegram* piece from March 19, 1896 (TFC).

Chapter VI

The epigraph is from the Tesla Memory Project (TMP).

I could find only one reference to Tesla's friendship with Sam Clemens in the writer's multivolume *The Autobiography of Mark Twain* (Los Angeles: University of California Press, 2010): a note on his alternating motor in Volume I, page 495. This is the source of the Twain quote on page 121.

The bedrock for Tesla's experiences in Colorado Springs is from his *Nikola Tesla: Colorado Springs Notes 1899-1900* (New York: BN Publishing, 2007). Unless you're an electrical engineer or someone versed in high-frequency electrical experiments, it's hard to decipher. The quote is from his journal. A more accessible description of his activities of that time is in "The Transmission of Electric Energy Without Wires," published in *Electrical World and Engineer*, March 5, 1904 (TFC). Seifer and Cheney have even better accounts, which I referenced.

On Wardenclyffe, I found archivist Leland Anderson's "Rare Notes from Tesla on Wardenclyffe" (http://www.tuks.nl/wiki/index.php/Main/TeslaRareNotesOnWardenclyffe) to be helpful, with technical details on what Tesla was building there. Also see Joe Sikorski's fine documentary *Tower for the People: Tesla's Dream at Wardenclyffe Continues* (http://www.imdb.com/title/tt3685200/). I met and interviewed Joe several times about the backstory of Wardenclyffe. The Tesla Science Center at Wardenclyffe site (http://www.teslasciencecenter.org/stanford-white/) was one of my first sources for information on Stanford White and the complex on Long Island.

Another important Tesla article I consulted was his "Tuned Lightning" piece from *English Mechanic and World of Science*, March 8, 1907 (TFC), in which he refers to "the wireless telephone" and introduces the idea of "stationary terrestrial waves" and the earth's "resonant vibration." These are two of the key concepts of telegeodynamics. An earlier Tesla piece referring to his experiments in Colorado Springs was "The Transmission of Electrical Energy Without Wires," which appeared in *Electrical World and Engineer*, March 6, 1904 (TFC).

The Morgan Library (ML) collection included well-known pieces penned by Tesla, including "The Wonder World to Be Created by Electricity" from *Manufacturer's Record*, Sept. 9, 1915, in which he makes an early reference to an "electric gun" and other innovations. The 1904 letter to Westinghouse about electric cars was in the Westinghouse archives (HHC).

Information about Tesla's proposed equation for modeling total human energy comes from his article "The Problem of Increasing Human Energy," which appeared in the June 1900 edition of *The Century Magazine*. Quotations from Tesla's autobiography (MI) also present some of his loftier visions.

There are few better writers and educators on relativity theory and modern physics than Columbia University professor Brian Greene. I consulted his *Smithsonian Magazine* piece "Gravity's Muse" (Oct. 2015) for information on relativity. Generally, any presentation or writing by Greene, who is ubiquitous on television and in print, will give you the skinny on contemporary thought in physics. I also saw Greene do a presentation at the Evestnet conference in Chicago on May 5, 2015. It was one of the best lectures on physics I've ever seen. Dennis Overbye's insightful "Finding Relativity" in the Nov. 24, 2015, *New York Times* was also helpful.

Chapter VII

The epigraph is from the TMP.

Erik Larson details the Tesla-Marconi narrative expertly in his *Thunderstruck* (New York: Random House, 2006).

Tesla's correspondence and repeated requests for money from the Morgans are documented in the letter series I acquired from the Morgan Library (ML). I also acquired the series of letters between J.P. Morgan and his son "Jack" from mid-1913 into 1916.

First-person accounts of Tesla's brushes with death, nervous breakdowns, and intense determination to recover and continue working are from his autobiography (MI).

I relied upon Seifer's account of the Wardenclyffe years for the story of Tesla's experience on Long Island.

For the TeslAction, I found some compelling words on creativity in Christie Aschwanden's "The Blessed Mess of Creativity," in the *New York Times*, Feb. 9. 2016.

Chapter VIII

The epigraph is from the TMP.

My description of the start-up of the Fisk turbogenerator and Insull's experiences are from MOP, McDonald's *Insull*, and Insull's *Memoirs.*

I relied upon the post-Wardenclyffe narratives from Cheney and Seifer. I also consulted Tesla's accounts of that time in MI.

Although I don't know if Julius Rosenwald ever met Tesla, he probably knew about his show at the Chicago World's Fair. I derived background on the philanthropist from Peter Ascoli's biography *Julius Rosenwald* (Bloomington: Indiana University Press, 2006).

For information on physicist Albert Michelson, I consulted this short biography: http://www.lib.uchicago.edu/projects/centcat/centcats/fac/facch07_01.html. There's also background on him on the Nobel site: http://www.nobelprize.org/nobel_prizes/physics/laureates/1907/michelson-bio.html.

For more on the University of Chicago's history, I consulted John Boyer's *The University of Chicago: A History* (Chicago: University of Chicago Press, 2015). I also heard Prof. Boyer lecture on the university's history several times in late 2015 on its Hyde Park Campus.

I referred to this short history of KDKA: http://ethw.org/KDKA,_First_Commercial_Radio_Station.

The last letters to Morgan and the Westinghouse Company were in HHC and ML archives.

Here's the original link for the Edison essay on Paine: http://learning.hccs.edu/faculty/jennifer.vacca/engl2327/author-presentations/the-philosophy-of-thomas-paine-by-thomas-edison-american-inventor-1925.

Christopher Nolan's 2006 film *The Prestige* (http://www.imdb.com/title/tt0482571/?ref_=nv_sr_1) is a must-see for Bowie's performance as Tesla. The book it was based upon by Christopher Priest has the same title (New York: Tor, 1995).

There is more on David Bowie in the *New York Times* piece by rock critic Jon Pareles, Jan. 11, 2016 (http://www.nytimes.com/2016/01/12/arts/music/david-bowie-dies-at-69.html?smtyp=cur&_r=0) and in "12 Things You Didn't Know About David Bowie" by Elsa Vulliamy in the British newspaper *The Independent*, also on Jan. 11, 2016. I saw the highly attended exhibit *David Bowie Is* at the Chicago Museum of Contemporary Art on Dec. 29, 2014. For more on this incredible show, see the companion book with the same title, written by Victoria Broackes (London: Victoria & Albert Museum, 2013). A short biography and interpretation of Bowie's early works can be found in *The Man Who Sold the World* (New York: Harper Collins, 2012) by Peter Doggett.

Tesla's Westinghouse letters from the 1930s were from HHC, although they appear to have been copied from the Tesla Museum.

The figures on sales of consumer appliances and account of that era were from David Nye's *Electrifying America: Social Meanings of a New Technology, 1880-1940* (Cambridge, Mass.: MIT Press, 1990), pp. 264–267.

Details on Tesla's flying machine were in George Trinkaus's *Tesla: The Lost Inventions* (Portland, OR: High Voltage Press, 1988), which also describes Tesla's bladeless turbine, coil, and magnifying transmitters.

Key missives and articles from Tesla in the 1930s were contained in "Tesla on Marconi's Feat," *New York World*, April 13, 1930; "Tesla Cosmic Ray Motor May Transmit Power 'Round the Earth," by John O'Neill, *Brooklyn Eagle*, July 10, 1932; "Man's Greatest Achievement," *New York*

American, July 7, 1930; "Our Future Motive Power," *Everyday Science and Mechanics*, Dec. 1931; "Tesla, Sure Life Exists on Other Planets, Works On at 76 to Establish His Belief," by William Engle, *New York World-Telegram*, July 9, 1932; "His Greatest Achievement," *New York Times*, July 11, 1935; "Nikola Tesla, At 79, Uses Earth to Transmit Signals, Expects to Have $100 Million Within Two Years," by Earl Sparling, *New York World-Telegram*, July 11, 1935; and "A Machine to End War," *Liberty* (as told to George Viereck), Feb. 1937.

I also referred extensively to Thomas Valone's *Harnessing the Wheelwork of Nature: Tesla's Science of Energy* (Kempton, Ill.: Adventures Unlimited Press), which examines Tesla's technologies in detail.

All the Tesla/Insull letters were from the TM, observed during an online document viewing on Feb. 18, 2016.

The inflation-adjusted estimate of Insull's contributions and average salary paid to Tesla in 1930 was from the U.S. Bureau of Labor Statistics (http://www.bls.gov/data/inflation_calculator.htm; http://www.bls.gov/opub/uscs/1934-36.pdf; http://www.simplyhired.com/salaries-k-1930-jobs.html).

Tesla's "death ray" pronouncements were reported on personally by John O'Neill and included in his *Prodigal Genius*, the first Tesla biography.

Chapter IX

The epigraph is from the TMP.

Terbo has offered accounts of his meeting with Tesla many times over the years. I interviewed him on Feb. 24, Nov. 12, and Nov. 19 in 2015. He sent me an afterword he'd written on Tesla titled "Extravagant Genius—Tender Uncle," which also recounts the meeting. He generously reviewed several sections of this book and offered corrections in handwritten notes and emails.

The last days of Tesla, particularly his meeting with King Peter II, are well documented in Cheney. The king's visit is also documented in the archives of the Hotel New Yorker, which printed up summaries of Tesla's stay there and maintains a "mini-museum" in its lower level (next to the business center). I also gathered material from a Tesla conference held there on Jan. 6, 2013, where I heard Matt Inman, Joe Kinney, and Jane Alcorn speak, among others cited in the text.

Tesla's letters and telegrams from the late 1930s and early 1940s were found in the Westinghouse archives (HHC), including his letter about raising chickens (May 22, 1941). The 1941 telegrams to Kosanovich were found in *Nikola Tesla: Correspondence With Relatives* (New York: Tesla Memorial Society, 1995), translated by Nicholas Kosanovich with the permission of the Tesla Museum, Belgrade. The collection, including letters between Tesla and William Terbo's father Nicholas, documents the inventor's money problems from the late 1920s to the end of his life.

Orson Welles's life is examined in Simon Callow's *Orson Welles: The Road to Xanadu* (New York: Viking, 1995). Also see http://www.history.com/this-day-in-history/welles-scares-nation.

I found the Einstein letter to FDR in Dan Cooper's *Enrico Fermi and the Revolutions of Modern Physics* (New York: Oxford University Press, 1999), which also provided background on the first sustained nuclear reaction at the University of Chicago. For more on the university's history, see http://www.uchicago.edu/about/history/.

I attended the Tesla exhibit on Chicago's Navy Pier and interviewed Vladimir Jelenkovic on Aug. 9, 2011.

One of my last FOIA requests, to the Defense Advanced Research Projects Agency (DARPA), a "component" of the Department of Defense, "located no documents" relating to Tesla (Feb. 16, 2012).

My NARA letter was from June 6, 2012.

The most heavily redacted FBI memorandum, dated Jan. 20, 1951, mentioned Spanel, Kosanovich, and "Colonel Erskine of military intelligence" (no specific branch cited). It was sent to Hoover's deputy, Clyde Tolson. I didn't see Tolson's specific reply in the FBI files, nor a clean version of the letter identifying the other parties involved.

I largely followed Seifer's narrative on the Tesla paper trail in his *Wizard* and his published dissertation *Nikola Tesla: Psychohistory of a Forgotten Inventor, Volumes I & II* (Ann Arbor, Mich.: UMI, 1987). I can't come close to duplicating the volume and quality of work Seifer has done in researching Tesla, his psyche, and the disposition of his papers.

The most ludicrous document I received from the FBI files was a memo dated June 7, 1947. It's so heavily redacted, the page is virtually all blacked out except for "Chicago" and some pronouns and verbs.

A notice on the U.S. Navy's new laser weapon appeared in "Navy Brings Out Futuristic Guns," by David Sharp, *Providence Journal*, Feb. 2, 2014.

The TeslAction on Tesla's loneliness was inspired by a talk given by Prof. John Cacioppo at Chicago's Wit Hotel on Oct. 29, 2015, as part of the University of Chicago's Harper Lectures series. I also referred to his book *Loneliness: Human Nature and the Need for Social Connection* (New York: Norton, 2008), which was coauthored with William Patrick.

Chapter X

The epigraph is from the TMP.

All my Philadelphia interviews were conducted at the Tesla People's Conference, sponsored by the Tesla Science Foundation, on July 10 and 11, 2015, at Independence Mall and at the Ethical Society of Philadelphia on Rittenhouse Square (http://teslasciencefoundation.com/). Nick Lonchar provided background by email.

The Tesla Conference 2011, sponsored by the Tesla Science Center at Wardenclyffe, was held at the Hilton Garden Inn in Riverhead, New York, on Nov. 5, 2011 (http://www.teslasciencecenter.org/2011/10/tesla-conference-2011-at-hilton-garden-inn-riverhead/). I spoke in the afternoon and interviewed Alcorn that day and in subsequent emails.

My discussion of Elon Musk relied largely upon Ashlee Vance's fine biography *Elon Musk: Tesla, SpaceX and the Quest for a Fantastic Future* (New York: Harper Collins, 2015). Drake Baer's "The Making of Tesla: Invention, Betrayal and the Birth of the Roadster," *Business Insider*, Nov. 11, 2014 (http://www.businessinsider.com/tesla-the-origin-story-2014-10), also provided some insights into Musk and the creators of the Tesla car:

I also referenced the blog and other background on Tesla Motors and its Powerwall system at https://www.teslamotors.com/.

For more on novel ways of creating rocket propulsion, see the NASA paper *Asymmetrical Capacitors for Propulsion* (NASA CR-2004-213312) and Paul LaViolette's *Secrets of Antigravity Propulsion: Tesla, UFOs and Classified Aerospace Technology* (Rochester, Vt.: Bear & Company, 2008). I can't vouch for the technological veracity of this book, but it's a place to start.

Quotations from Tesla about some of his more unusual ideas are found in his autobiography (MI).

Although Larry Page rarely talks to the media, his role at Google/Alphabet is profiled in Ken Auletta's *Googled: The End of the World As We Know It* (New York: Penguin, 2009) and an excellent piece titled "The Unknown Story of Larry Page's Comeback" by Nicholas Carlson, *Business Insider*, April 24, 2014 (http://www.businessinsider.com/larry-page-the-untold-story-2014-4).

Also helpful was Chade-Meng Tan's "Scaling Compassion: The Story of Google Employee #107," interview by Karen Christensen in the *Rotman Management Magazine*, Winter 2016, pp. 43–47.

I saw Walter Isaacson lecture at the Chicago Humanities Festival on Sept. 15, 2015, at the Fourth Presbyterian Church in Chicago and condensed the main points of his observations. I also referenced his book *The Innovators: How a Group of Hackers, Geniuses and Geeks Created the Digital Revolution* (New York: Simon & Schuster, 2014).

The "Fourth Industrial Revolution," championed by Klaus Schwab, founder of the World Economic Forum, is a meta-integration of multiple trends in technology and life sciences (http://www.weforum.org/pages/the-fourth-industrial-revolution-by-klaus-schwab).

Some futurists, such as Jeremy Rifkin, claim that we're still in the "third" industrial revolution. See his *The Third Industrial Revolution: How Lateral Power Is Transforming Energy, the Economy and the World* (New York: Palgrave, 2011). I won't summarize his observations here, but Rifkin asserts that the Internet and other technologies are making energy cheaper, among other things. Both Schwab and Rifkin are well worth reading, although they differ in terminology.

On wireless buses, see "South Korea tests first 'wireless road'" (http://america.aljazeera.com/articles/2013/8/12/south-korea-develops worldsfirstelectricroad.html).

Upon my last visit to New York City, I saw Daniel Doctoroff speak on Oct. 8, 2015, at the Society for American Business Editors and Writers (SABEW) fall conference at the CUNY campus in midtown Manhattan. The same day, I visited Tesla's room nearby in the Hotel New Yorker.

Nicholas Epley, a professor at the University of Chicago, lectured in Chicago at the University's Gleacher Center Oct. 15, 2015. I also referenced his book *Mindwise: How We Understand What Others Think, Believe, Feel and Do* (New York: Knopf, 2014).

On space-based solar power, I attended a conference on May 27, 2010, in Chicago. It was titled "Four Decades After Apollo: Getting Back to the Future" and was sponsored by the National Space Society.

The Microgrid is installed at the Illinois Institute of Technology's Galvin Center for Electricity Innovation on Chicago's South Side, right across from Comisky Park, where the White Sox play. See http://www.iitmicrogrid.net/.

For more on Kurzweil's "singularity," see http://www.singularity.com/.

I saw Chungliang Al Huang at Common Ground in Deerfield, Illinois, in 2014. I quoted from his book *Quantum Soup: Fortune Cookies in Crisis* (London: Singing Dragon, 2011).

Epilogue

Hawking's pronouncement was in "Hawking: Humans at Risk of Lethal "Own Goal," BBC.com, Jan. 19, 2016.

You can find *Laudato si* at www.vatican.org. It's well worth reading, even if you're not Catholic.

The World Economic Forum was the prime source for a number of trends, including job loss and robotics. See its "Deep Shift" report at http://www3.weforum.org/docs/WEF_GAC15_Technological_Tipping_Points_report_2015.pdf. The WEF was also the source for the Shiller quote.

For insights into robotics and warfare, see *Wired for War* by P.W. Singer (New York: Penguin, 2009). Although the numbers are dated, it gives a history of how far robotics has advanced in battlefield applications.

On the wireless energy front, check out the Russian project globalenergytransmission.com, which hopes to take up Tesla's dream of broadcasting wireless power.

I interviewed Victor Zaderej several times in late 2015 and exchanged emails with him. Also see his piece on his passive home (http://www.homepower.com/articles/home-efficiency/design-construction/american-passive-home) and his TEDx talk on LED lighting systems (https://www.youtube.com/watch?v=XR2Ihxu7HJ4).

Although you could devote volumes to the Maker Movement and its Maker Faires, a good resource is http://makerfaire.com/maker-movement/.

Paul Krugman discussed the politics and economics of climate change in his Feb. 1, 2016, *New York Times* column "Wind, Sun and Fire" (http://www.nytimes.com/2016/02/01/opinion/wind-sun-and-fire.html?_r=0.)

See Jeremy Rifkin's rebuttal to the WEF's "Fourth Industrial Revolution" platform at http://www.huffingtonpost.com/jeremy-rifkin/the-2016-world-economic-f_b_8975326.html.

The Freud quotation comes from *Civilization and Its Discontents*, trans. James Strachey (New York: W.W. Norton, 1962), p. 49.

ACKNOWLEDGMENTS

This book would not have been possible without the generous assistance of people from across the world. So many helped in myriad ways—forgive me if I've neglected to mention you. I will not forget your aid.

William Terbo, Tesla's grandnephew, was extremely generous with his time and voluminous archives. Jane Alcorn at the Tesla Science Center at Wardenclyffe was invaluable in walking me through the current effort to restore Tesla's lab on Long Island (http://www.teslasciencecenter.org/). Also, the Long Island filmmakers Joseph Sikorski and Michael Calomino, who have worked tirelessly on their Tesla film *Fragments from Olympus* and their terrific documentary *Tower to the People*, lent their insight on the inventor. At the Wyndham Hotel New Yorker, thanks to chief engineer Joe Kinney, the resident Tesla archivist and excellent tour guide.

Nikola Lonchar, founder/president of the Tesla Inventor's Club/Science Foundation in Philadelphia, was of great assistance, as were Sam Mason, Sherry Kumar, David Vuich, Marina Schwabic, Zvezdana Stojanovic Scott, Ashley Redfearn, Tim Eaton, James Jaeger, Brian Yetzer, and Zeljko Mirkovic.

In Chicago, thanks to Dr. Milena Bajich, leading light of the Chicago Tesla Club, and to the Serbian American Museum St. Sava, particularly Vesna Noble and Dr. Zivojin Pavlovic (http://www.serbianamericanmuseum.org/contact-us/).

The idea for this book was seeded by a single letter from the Samuel Insull papers at Loyola University Chicago. The original archivist I worked with was Kathy Young more than a decade ago.

I received generous assistance from the librarians and archivists at the Sen. John Heinz History Center in Pittsburgh, particularly Carly Lough and Liz Simpson in the library.

My many friends in Belgrade at the Tesla Museum were instrumental in patiently guiding me through thousands of pages of documents over a period of several months. Heartfelt thanks to Branimir Jovanovic, Vladimir Jelenkovic, Radmila Adzic, Ivana Ciric, and Milica Kesler. Mr. Jelenkovic was extremely generous, not only granting me a personal interview, but providing me with copies of Tesla's patents, his autobiography, and several other Tesla writings.

No chronicler of Tesla's life and time can stand alone. I'm indebted to the fine work already accomplished by Marc Seifer, Margaret Cheney, W. Bernard Carlson, David Kent, and John O'Neill. They laid the groundwork for what we know about Tesla.

A particularly hearty thanks to the many reference librarians at the Grayslake Area Public Library in Grayslake, Illinois, who have endured my many obscure requests for years, led by my friend Roberta Thomas. You're the warm, beating heart of free knowledge.

For my literary agent, Marilyn Allen, and my former agent, Robert Shepard, who both helped usher this book into the world, thanks for your direction, persistence, and patience. And speaking of patience, I appreciate the kind attention, diligence, and masterful production of my Sterling Publishing editor Melanie Madden and her many colleagues, as well as Lary Rosenblatt, Laurie Lieb, and Alan Barnett at 22MediaWorks. They worked tirelessly to produce this book.

Certainly, all the Tesla organizations and supporters from Australia to Europe had a part in this book. While I could not mention or cite your work individually, many thanks.

And finally, to my wife Kathleen and daughters Sarah and Julia and all my friends and neighbors, who have wondered if I'd ever finish this book, my love to you all.

IMAGE CREDITS

akg-images
© TT News Agency/SVT: 22

Alamy
© The Advertising Archives: 199; © Everett Collection Inc: 203; © Glasshouse Images: 196; © Dennis Hallinan: 31; © Len Holsborg: 218; © Mary Evans Picture Library: 43; © North Wind Picture Archives: 81; © World History Archive: 143

Art Resource, NY
The Morgan Library & Museum: 169

Courtesy Dover Publications
32

Galvin Center for Electricity Innovation
231

Getty Images
© New York Daily News: 178; © Chris Walter: 175

Heinz History Center
106 top

IEEE History Center
Images courtesy ethw.org/Archives:Papers_of_Nikola_Tesla: 152-153, 194

Courtesy Joe Kinney/New Yorker Hotel
8, 11, 188, 191, 216

Library of Congress
4, 10, 25, 36, 76-77, 89, 97, 100 101, 102, 109, 147

NASA
viii

National Archives
177

old-photographs.com
160

Private Collection
xii, 2, 13, 14, 20, 51, 62, 72, 85, 88, 90, 94, 96, 108, 111, 113 left, 123, 126, 127, 128, 140, 145, 148, 165, 166, 167, 171, 172, 180, 181, 186, 204, 210

Shutterstock
© Hadrian: 222; © Asif Islam: 3, 225; © trekandshoot: 212

Smithsonian
National Museum of American History: 78, 113 right, 118, 125

The Tesla Collection
144

Tesla Wardenclyffe Project
24, 98, 130

US Patent and Trademark Office
33, 64, 133

Wellcome Library, London
ii, 40, 68

Courtesy Wikimedia Foundation
28, 41, 45, 48, 54 top, 56, 63, 65, 71, 86, 103, 104, 114, 135, 158, 160; Andreas F. Borchert: 238; Boston Public Library Tichnor Brothers Collection: 54 bottom; Gilder Lehrman Collection: 75; Internet Archive: 106 bottom, 120; ISA Internationales Stadtbauatelier: 233; Melvin A. Miller of the Argonne National Laboratory: 192; Museum of Innovation and Science, Schenectady, NY: 79; Pearson Scott Foresman: 6; Camilo Sanchez: 224; Southern Methodist University: 73; SpaceX: 220; Tamorlan: 70; Lucas Taylor/CERN: 206

© Yetzer Studio
interactive color inserts

INDEX

absolute zero, 42
AC technology
Chicago plan, 158, 160
development, 1–2, 83–84
embodying power of, 97
energizing the world, 10
equipment, 91
in Germany, 105
hyper-connected consumer culture, 197
motor, 70, 73, 83–85
national power system, 122
Niagara Falls dynamos, 106–8
perfecting, 87
power, 159
understanding, 78
vision of, 90
Adamic, Louis, 194
Adams, Henry, 107
Addams, Jane, 102
advocacy, 155, 184
air conditioning, 162
Al Huang, Chungliang, 236
Alcorn, Jane, 217
Alphabet Inc., 3, 225–26, 229
alternative energy, 242
Amazing Stories, 186
ambiverts, 58
American Institute of Electrical Engineers (AIEE), 84, 152–53
American Philosophical Society, 227
Anderson, Leland, 164
animal electricity, 42
Anthony, Susan B., 102
anti-Semitism, 173
antigravity, 213
apparatus for producing ozone, 133
Apple, 116, 157, 225, 227, 228
Arab sailors, 29
arc lighting, 74, 82–83
artificial intelligence, 230, 240–42
Astor, John Jacob, 122, 128, 148
asymmetric capacitors, 213–14
autobiography, 166
automobiles, 103
autotelic self, 60
Aviation Week and Space Technology, 21

Babbage, Charles, 116
Bajich, Milena Tatic, 16, 57–60
Batchelor, Charles, 72, 73, 75, 80
Battle of Currents, 86–89, 104, 198
Battle of Kosovo, 56
Baum, Frank L., 102
Belefski, Carolyn, 234
Bell telephone network, 146
Bennett, Edward, 158, 160
Bezos, Jeff, 184
big data, 229, 240
birthday celebrations, 12
bladeless turbine, 146, 149, 155, 166–67
Bloom, Harold, 36
Bloomberg, Michael, 229
blowers, 168
Bohr, Niels, 208
Boldt, George, 142, 162
boom-and-bust saga, 200
Bowie, David, 174–76
Branson, Richard, 184
Brin, Sergey, 225
Brisbane, Arthur, 59
Brooklyn Eagle, 180
Brown, Alfred, 83–85, 96–97
Brown, Harold, 86, 116
Broz, Josip, 192
Budapest, Hungary, 62, 71–72
Buffet, Warren, 178
Burnham, Daniel, 98–99, 158, 160
business repositioning, 168–73
Byron, George Gordon, 44–45
Cacioppo, John, 208
cameras, 103
Capone, Al, 178–79
Capra, Fritjof, 34
Carabeo, Joe, 234
carbon-free energy, 205
Carlson, W. Bernard, 14, 53–57, 63, 66, 117, 203
carousing, 58
Carrier, Willis, 162
Carroll, Liz, 239
celebrity, 5
celibacy, 121
Century Magazine, 109, 111, 121, 127
Chicago
AC technology plan, 158, 160
Haymarket Square, 102
luminous modern city, 54
as Oz, 102
Serbian population, 200as
toxic miasma, 99
Chicago World's Fair
Electricity Building, 103
inventions debuted, 103
Museum of Science and Industry, 104
Parliament of World's Religions, 102
Westinghouse Electric Company at, 98
as White City, 98–99, 100–101
wizard of electricity at, 95–96
childhood, 49–50
chimera, 5–6, 236–37, 240
cholera, 110
civic groups, 209
Civilization and Its Discontents (Freud), 243
Civric, Zorica, 29
Clark, George, 193
climate catastrophe, 242
climate change, 240
clock assembly, 30

CMS particle detector, 206
Codex Atlanticus, 30
cogeneration, 132
Cold War, 20
collaboration, 228
Collard, Hippolyte-Auguste, 73
college loan debt, 229
Colorado Springs lab, 122–27, 146–47
comfort zone, 47
Common Sense (Paine), 170
communication, 75
communism, 170, 193
compassion, 189, 243
compressors, 168
concentration, 67
conspicuous consumption, 200
Consumer Reports, 222
contemplation, 57
convergent thinkers, 57–58
core qualities, 27
Corning Glass, 228
cosmology, 24
creativity
  ambiverts and, 58
  applying, 240
  convergent and divergent thinkers, 57–58
  defying stereotyping, 59
  diligence and, 58
  fantasy and, 58
  flow driving, 60, 63
  humility tempering, 59
  intrepid, 3, 5–6
  intuitive approach to, 56
  merger of art and science, 227
  in notes and diagrams, 23
  reboot creative process, 7
  suffering and, 59
Csikszentmihalyi, Mihaly, 60
curiosity, 46–47
Curl, Robert, 69
cybernetics, 230

da Vinci, Leonardo
  Codex Atlanticus, 30
  curiosity of, 12
  flying machine, 32
  grip of, 46
  hydraulic devices, 31
  interests, 29–30
  virtuoso, 227
  water flow study, 34
Daily Calumet, 200
dancing to-loop alternator, 63
Darrow, Clarence, 102
Darwin, Charles, 51
Davos Forum, 241–42
Davy, Humphrey, 39, 74
DC technology
  motor, 61
  power, 75, 80, 81
  selling, 80
death, 193
death ray
  announcement, 11–12
  development, 5
  drawings, 202
  letters, 201
  neutron-based weapon, 190–91
  press dubbing, 183
  rough drawings, 12
Defense Advanced Research Projects Agency (DARPA), 21, 205
Department of Peace, 241
depression, 15
Descartes, René, 40, 41
descriptive statistics, 51
digitalization, 243
diligence, 58
disinfection, 132
ditch digging, 83, 116
divergent thinkers, 57–58
Divina, Mano, 214
Doctoroff, Daniel, 229
Dollard, Eric, 232
Donne, John, 209
Dracula (Stoker), 44
Dreiser, Theodore, 102
drones, 3, 5, 114, 133, 171, 241
Dvorak, Antonin, 102
dynamos
  Gramme, 69, 70
  high-frequency, 168
  Niagara Falls, 106–8

Eberhard, Martin, 219, 221
eccentricities, 12–13
Edison, Thomas
  archives, 201
  Battle of Currents, 86–89, 104, 198
  brand value, 74
  celebrity, 23, 170
  central power station, 1–2
  fine-tuning projects, 6
  Ford and, 105, 170
  gas and, 74
  as hero, 65
  hiring repairman, 79–80
  humiliation from, 82, 91
  Insull departing, 161
  lightbulbs and, 80, 198, 201
  as nemesis, 5
  opposition of, 59
  as overshadowing, 217
  Pearl Street Station, 79
  planned empire, 80
  plastic approach, 56
  as rival, 24
  with wax cylinder phonograph, 79
  workaholic, 78
Edison General Electric, 88, 159
Edison Machine Works, 72, 74–82
education, 40, 50–51, 69
effete demeanor and dress, 59
Eggs of Columbus, 95–96, 117, 201
eidetic memory, 56
Einstein, Albert
  biography, 209, 227
  grand unification theory, 138
  on nuclear chain reaction, 197
  publication, 134–36
  on speed of light, 182
  theory, 205
  view of infinite, 174
  violin and, 227
electric chair, 86–87
electric propulsion, 133
Electrical Club New York, 84
electrical engineer, 6
Electrical Experimenter, 127, 164, 166, 204
Electrical World and Engineer, 143, 144
electricity. *See also* AC technology; DC technology; wizard of electricity
  channeling, 127
  electrical grid, 1–2
  electrification of homes, 173
  in Gilded Age, 82
  global super highway, 138
  infrastructure, 172
  new electrical age, 197–200
  one-lever control, 68
  Roosevelt electrification, 132
  wireless power, 231
Electrifying America (Nye), 198
electrochemistry, 39
electrogravitics, 214
electromagnetism, 39

electrostatic generator, 13
empathy, 189
energy. *See also* AC technology
  alternative, 242
  carbon-free, 205
  free, 182
  human, 114, 127, 134
  hydrodynamic, 10
  oscillating, 5, 112–13, 168
  radiant, 87, 182
  solar, 134, 242
  vision of, 5
  zero, 215, 242
Energy Independence Celebration, 2015, 211–13
engagement, 209
epiphanies, 66
Epley, Nicholas, 232–33
eponymous coil, 87
ethics, 234
eugenics, 51
eureka moment, 34–35
Exelon Corporation, 2
extraterrestrial life, 182

failure, 154–56
faith, 234
Fannie Mae, 154
fantasy, 54, 58–59
Faraday, Michael
  career, 42
  Davy as mentor, 74
  experimentation, 39
  grip of, 46
  magnetic sparking coil, 40
  transatlantic cable, 44
fascism, 170, 193, 239
Faust (Goethe), 45, 50, 56, 61–63
Federal Bureau of Investigation (FBI)
  catching attention of, 115
  files, 65, 91
  memos, 21, 202, 205
  monitoring, 9–10
  papers confiscated by, 19
  requests to, 20
feeble forces, 35
Fermi, Enrico, 193, 208
Ferraris, Galileo, 84
financial collapse, 23
Firestone, Harvey, 170
Fitzgerald, Bloyce, 21
flow, 60, 63
fluid diode, 164
flying machine, 32, 33
Ford, Henry
  anti-Semitism, 173
  Edison and, 105, 170
foreign agents, 9
fortitude, 154–55
fossil fuels, 134, 213, 232
fountains, 168
Fourth Industrial Revolution, 227–31, 240
Francis, Pope, 241
Frankenstein, or The New Prometheus (Shelley, M.), 44
Franklin, Benjamin
  biography, 209, 227
  capturing lightening, 46
  electrical experiments, 37–38
  lightning rods, 211
  portrait, 28
Freddie Mac, 154
free energy, 182
Freedom-of-Information Act, 19, 21, 154, 200
Freud, Sigmund, 243
funeral, 23

Gaia Theory, 34
Galileo, 136
Galton, Francis, 51
Galvani, Luigi, 42
Galveston Daily News, 181
gambling, 58
Gardner, Howard, 236
gas turbines, 168
Gates, Bill, 178
General Electric (GE), 91, 105, 122, 143, 170, 198–99
generators. *See also* dynamos
  electrical, 216
  electrostatic, 13
  turbogenerator, 159–62
genomics, 230, 240
Gernsback, Hugo, 164
The Girl in the Red Velvet Swing (film), 146
Gladwell, Malcolm, 92
global warming, 240, 241
Goethe, Johann Wolfgang von, 44–45, 50, 56, 61–63, 70
Google, 3, 224–27, 241
Gould, Jay, 83
Gracac village, 54
Gramme dynamo, 69, 70
grand unification theory, 138
Grant, Ulysses, 111
gravity, 35, 136, 174, 197
Graz Polytechnic, 69–71
Great Depression, 2, 176–78, 180, 188

hallucinations, 52
Hardy, Thomas, 184
Hawking, Stephen, 240
Heisenberg, Werner, 208
helicopter, 173
Herr, E. M., 167–68
Higgs boson, 205
high-frequency dynamos, 168
Hitler, Adolf, 10, 12, 51, 182, 190–91
holistic legacy, 6
home illumination, 162
Hoover, J. Edgar
  assistants, 202
  disclaimer, 20
  as paranoid, 10
  spying by, 19
  with Tolson, 203
  on weapons, 193
Hotel New Yorker
  conferences, 189, 201, 221
  electrical generator, 216
  final home, 8–10
  last visit, 234–35
  modest rooms, 188, 208
  in 1930s, 22
  for tech mavens, 215
Houseman, John, 195
hubs, 93
human energy, 114, 127, 134
humanitarian, 136–37, 151
Hume, David, 40
humility
  creativity tempered by, 59
  tolerance and, 227
  of Twain, 121
Hunt, Samantha, 176
Hurricane Katrina, 5
hydrodynamics, 10, 31, 56
hydroelectric power, 89, 132
hyperloop transportation system, 223–24
hypoid spiral bevel gear, 189

idea-sharing, 209
imagination, 53, 55, 66
incandescent lighting, 56, 69, 75, 82
Inman, Matt, 217
innovator, 6, 227–28
The Innovators: How a Group of Hackers, Geniuses and Geeks Created the Digital Revolution (Isaacson), 227

Insull, Samuel
  on AC equipment, 91
  as confidant, 105
  conglomerate, 170
  departing from Edison and Morgan, J.P., 161
  downfall, 2
  at Edison Company, 159
  fraud trials, 179–80
  letter to, 1–2, 19, 201
  as manager, 75
  marketing maven, 80
  multitasker, 78
  stock capital, 143
  stock sales, 178–79
  support from, 177–78
  turbogenerator and, 161
integral ecology, 241
interference preventers, 168
interferometer, 164
Internet, 230, 240
interstate highway system, 197–98
introspection, 26–27
intuitive approach to creativity, 56
inventions
  Chicago World's Fair and, 103
  epoch-making, 180
  last, 173–74
  metaphor becoming, 63
  science and, 172
  Twain and, 24, 111, 121
The Invention of Everything Else (Hunt), 176
ionosphere, 124
Isaacson, Walter, 209, 227
isolation, 121, 192, 208–9

Jelenkovic, Vladimir, 200–201, 217
Jobs, Steve, 157, 209, 227, 228
Johnson, Katherine, 109, 121, 146, 162
Johnson, Robert Underwood, 109, 121, 146, 162
Joplin, Scott, 102
Josika, 18
The Jungle (Sinclair), 99

Karabeg, Dino, 211
Karlobag, Croatia, 48
KDKA, 167
Kemmler, William, 86–87
Kent, David, 91
Kinney, Joe, 214, 216–17
Kipling, Rudyard, 99
Kissinger, Henry, 209
Kosanovich, Sava, 11, 190–91, 193, 202
Krugman, Paul, 242
Ku Klux Klan (KKK), 173
Kumar, Sherry, 214
Kurzweil, Ray, 230, 237

Lane, Robert, 82–83
learning, 209
Lehman Brothers, 2, 154, 179
Lenin, Vladimir, 170
letterhead, 168–69
life experiences, 27
light flashes, 49
lightbulbs
  beyond, 3
  business, 74
  ceiling, 198
  creating, 228
  Edison and, 80, 198, 201
  illuminating, 110
  incandescent, 56, 69
  powering, 127
  wireless fluorescent, 213
lightning
  artificial, 122
  bolts, 5, 45
  capturing, 46, 107
  coils, 124
  power of, 44
  protectors, 168
  rods, 38, 95, 124, 211
  truth in, 242
Lindbergh, Charles, 174
Logan, Teresa Roberts, 234
logic gate, 164
Lonchar, Nick, 212–13
loneliness, 194
Lough Derg, 239–40
Lovelace, Ada, 45, 116
Lowenstein, Fritz, 124
luminosity, 237
luminous phenomena, 52–53
Lusitania, 162

Madison Square Garden demonstrations, 113
magic, 35–36, 117, 214
magnetic field
  Descartes drawing, 41
  magnetic sparking coil, 40
  rotating, 60–65, 70
magnetic resonance imaging (MRI), 208
magnifying transmitters, 113, 126, 189
Manhattan lab, 114, 119, 122
mania, 14, 18, 126
Manufacturer's Record, 131, 134
Marconi, Guglielmo
  charges of, 142–43
  early radio apparatus, 128–29
  patent suit, 87, 162, 195
  railing against, 180
  technology, 124, 128–29
  wireless transmitter schematics and, 146–47
Maribor, Slovenia, 71
Mason, Sam, 213–14
McCarthy, Joseph, 205
McKinley, William, 142
memory
  eidetic, 56
  loss, 121
  memorable experiences, 67
mental illness, 14–19
meta-integration, 6, 226, 230–32, 240–43
metaphor, becoming invention, 63
Michelangelo, 51, 53
Michelson, Albert, 164
Michelson-Morely experiment, 164
microgrids, 230–31
Microsoft, 225
migraine headaches, 52
mind-body connection, 24
mind journeys, 53, 66
Mindwise (Epley), 233
miracles, 65, 242
Mirkovic, Zeljko, 214
Mischel, Walter, 18
Morgan, Anne, 121, 124
Morgan, J. P.
  affiliates, 170
  backings, 83
  cartel, 74, 161
  as celebrated, 23
  circle of, 145
  consolidations, 1, 88, 115
  death of, 148–49
  grip on company, 91
  interests, 105
  investments, 59
  loans from, 124, 131
  Murray Hill mansion power, 81
  noble generosity, 143

owning power, 122
support from, 122
Morgan, Jack
appeal to, 169–70
cutoff letter from, 149
loan from, 148–49
Morris, Gene, 202
motors. *See also* Tesla Motors
AC technology, 70, 73, 83–85
DC technology, 61
Mount Tambora eruption, 44
Muir, John, 110
Museum of Science and Industry, 104, 163–64
Musk, Elon, 3, 184, 200, 218–19, 223–24, 231
Mussolini, Benito, 51, 182
Mutual Union, 83
The Mysterious Stranger (Twain), 120–21

National Archives and Record Administration, 202
National Electrical Light Association, 84
natural laws, 39
navigation, 133
Nazism, 171, 197, 239
nervous breakdowns, 130–31
networking, 157
New Agers, 5
New Deal, 179
new electrical age, 197–200
New York City power lines, 76–77
New York Times, 12, 155, 193
New York World, 97, 180
Newton, Isaac, 136
Niagara Falls power station
AC dynamos, 106–8
big wheel pictured, 34
engineer for, 59
hydrodynamic energy, 10
Westinghouse Electric Company and, 105–7
noble purpose, 115
Nolan, Christopher, 174
nuclear reactor, 192, 193, 207
nuclear war, 240
number three obsession, 10–11
Nunn, L. L., 89
Nye, David, 198

observations of nature, 35
obsessive-compulsive disorder (OCD), 16
Office of Alien Property (OAP), 9, 11–12, 202
Olmsted, Frederick Law, 98–99
On the Origin of Species (Darwin), 51
O'Neill, John, 23, 180, 224
Ophir Colorado hydroelectric plant, 89
Oregon, 80
organization, 75
oscillating energy, 5, 112–13, 168
ozone, 132–33
"Ozymandias" (Shelley, P.), 217

pacifism, 10
Page, Larry, 3, 224–26, 229
Paine, Thomas, 170
Paris, 72–73
particle beam experiments, 21, 206–7
Passio, Mark, 214
passion, 228
patents, 145, 162, 164, 173–74, 195
Paul, Frank R., 172
Pauli, Wolfgang, 208
pearl aversion, 17
Pearl Street Station
activating, 79
inefficiency, 81
opening, 107
tubes and wires, 81–82
Peck, Charles, 83–85, 96–97, 116
Person Magazine, 94
Peter II, 190–92, 240
pigeon obsession, 23–24, 190, 239
Poeschl, Jacob, 69–70
poetry, 24
Poldhu Wireless Station, England, 143
Polidori, John, 44–45
polyphase project, 87, 97–98
pop-gun manufacture, 30
Popular Science, 68, 164
poverty, 194
power. *See also* Niagara Falls power station; wireless power
AC technology, 97, 122, 159
DC technology, 75, 80, 81
Edison's central power station, 1–2
hydroelectric, 89, 132
of lightning, 44
Morgan owning, 122
Murray Hill mansion, 81
national system, 122
New York City lines, 76–77
towers of, 110
Westinghouse owning, 122
Prague, Czech Republic, 71
precision instruments, 168
The Prestige (film), 174–76
Priest, Christopher, 174
Prljevic, Mirjana, 159
"The Problem of Increasing Human Energy" (Tesla, N.), 127
Prometheus, 45, 242–43
prophecies
in array of innovations, 131–34
cogeneration, 132
disinfection, 132
electric propulsion, 133
for future of humanity, 181
hydroelectric power, 89, 132
navigation, 133
new view of time and space, 134–37
solar energy, 134
wireless weapons, 133
Protic, Aleksandar, 141, 217
public health, 128
public transportation, 162
punishment, 18

quality of life, 128, 243
quantum computers, 242

radiant energy, 87, 182
radiation fusillade, 126
radio
home use, 162
KDKA station, 167
mass media form, 168
popularity, 173
technology, 10, 129
transmitter, 114
Radio Corporation of America, 146
ragtime, 102
Reagan, Ronald, 21, 202–3
recluse, 23
Redfearn, Ashley, 214
regenerative entrepreneurs, 226
relativity, theory of, 134, 136, 205
religion, 24, 182
remote control, 3, 112–13

repositioning, 184–85
research project, 47
resilience, 149–51, 156–57
resonance, 114
Rifkin, Jeremy, 243
Roaring Twenties, 166–67, 170, 197
robotics, 3, 240–42
rockets. *See also* Space X
research, 10
reusable, 242
rocket, 213–14
Roman engineers, 29
Roosevelt, Eleanor, 192
Roosevelt, Franklin, Delano, 2, 132, 179
Roosevelt, Teddy, 142
Rosenwald, Julius, 163
rotating magnetic field, 60–65
routine, 151, 154
royalties, 173

safety razors, 103
San Francisco Call, 147
Sandburg, Carl, 102
Sargent, Fred, 161
Science and Invention, 172
Scientific American, 108
scientific novelties, 168
SciTech, 201
Second Industrial Revolution, 62, 162
Seifer, Marc, 11, 21, 162, 205–7, 217
self-reliance, 234
sensory experiences, 67
Sforza family, 30
Shakespeare, William, 35–36, 55
Shallenberger, Oliver, 84–85
Shelley, Mary, 44, 120
Shelley, Percy, 44, 217
Shiller, Robert, 241–42
short-termism, 139
Sidewalk Labs, 229
Silicon Valley, 84, 157, 219
Sinclair, Upton, 99
singularity, 237
The Singularity Is Near (Kurzweil), 230
Sistine Chapel, 51
smart city, 229, 233
Smith, M. W., 203
snowball avalanche, 35
social circle, 109–12
Society of American Business Editors and Writers, 227
solar energy, 134, 242
solitude, 157
soup kitchens, 177
Space X, 3, 219–21
Spanel, Abraham, 193, 202
speed of light, 164, 182, 205
speedometer, 164–65, 168, 169–70
"Spirit of the Earth," 65
spiritual economics, 243
standing or stationary waves, 124–25
steam turbines, 159, 168
Steinmetz, Charles Proteus, 122
stereotyping, 59
Stoker, Bram, 44
A Story of Youth Told by Age (Tesla, N.), 42
Strategic Defense Initiative (SDI), 21, 202–3
Straubel, J. B., 219
Stubenrauch, Hans, 65
suffering
creativity and, 59
maladies, 192
observation, 26
Sully Bridge, Paris, 73
superviruses, 240
survival, 150, 233
Swezey, Kenneth, 193, 202
synesthesia, 17
system building, 3
Szigeti, Anthony, 61, 150

Tarpenning, Marc, 219
teamwork, 228
techno-cultural shock wave, 243
telautomaton, 113
telegeodynamics, 12, 182, 203, 204, 214
telephone, 127, 146
telescoping antenna, 125
television, 127
The Tempest (Shakespeare), 35–36
Terbo, William, 187–90, 201, 217–18
terrestrial motions at a distance, 243
Tesla, Dane, 14–15, 26, 49–50, 240
Tesla, Djuka, 50, 51
Tesla: Inventor of the Electric Age (Carlson), 203
Tesla, Milutin, 17, 42, 49, 51, 71
Tesla, Nikola, 14, 22, 78, 109, 111, 123, 148, 180, 188, 191, 210. *See also specific topics*
Tesla: The Wizard of Electricity (Kent), 91
Tesla coil, 87, 122, 124, 213–14
Tesla Electric Light &
Manufacturing, 83, 115
Tesla Memory Project, 217
Tesla Motors, 3, 200, 219–22, 241
Tesla Science Foundation, 212–13, 217
TeslActions
about, 7
be chimeric, 236–37
be curious, 46–47
be emphatic, 208–9
be indispensable, 92–93
be introspective, 26–27
be resilient, 156–57
practice visualization, 66–67
reposition yourself, 184–85
sell your ideas, 116–17
zoom out, 138–39
Teslaphiles, 6, 201, 213, 214, 217
thermodynamic transformers, 132
Thomson, Elihu, 84
Thomson, William (Lord Kelvin)
absolute zero and, 42
conductor, 44
demonstrations, 43
enterprising, 40
Time magazine, 180, 188
Titanic, 148
Toffler, Alvin, 230
tolerance, 227
Tolson, Clyde, 202, 203
Tom Sawyer (Twain), 119
torpedoes, 21, 133
towers of power, 110
transcending mechanical sense, 42
Trbojevich, Nicholas, 189
Trump, John, 21, 193
trust-busting, 142
turbogenerator, 159–62
Turing, Alan, 116
Twain, Mark
as celebrated, 23
friendship of, 113
humility of, 121
investing in inventions, 24, 111, 121

as reviving, 110
with vacuum lamp, 111, 119–20
typewriters, 103

uber-machine age, 240
undisturbed thought, 55

vacuum lamp, 111, 119–20
vacuum pump, 168
vagrancy, 71
Vail, Benjamin, 82–83
The Vampyre (Polidori), 44–45
Varosliget Park, Budapest, 62
Veblen, Thorstein, 200
Vedas, 24
Venturi, Robert, 211
Verne, Jules, 159
vertical takeoff and landing vehicle (VTOL), 173–74, 214
Villard, Henry, 88
vision
of AC technology, 90
creative, 29
of energy, science, and world peace, 5
guiding, 138
unusual, 150
as working models, 58
World System, 185
visualization
abilities, 61–62
practicing, 66–67
structured, 53, 55–56
Vivekananda, 24, 102, 114
Volta, Alessandro, 42
Voltaic pile, 42
Voltaire, 69, 70
volunteering, 157
Vukasinovic, Nevena, 187
Vunjak-Novakovic, Gordana, 119

Waldorf-Astoria Hotel
high-style living, 24, 128
as residence, 122
socializing at, 109
Waltham Watch Company, 164–65, 168, 169–70
War of the Worlds (Wells, H. G.), 186, 195–97
Wardenclyffe Tower lab
abandonment, 154–56
decaying, 217
as electronic shield, 162
fall of, 140–46
logo, 168
photo, 118
plans and building, 127–31
revival attempt, 167
saving, 215–17
virtual tour, 217
what might have been, 163
Washington Herald, 4
water flow study, 34
Waterside Station, 146
wax cylinder phonograph, 79
weapons. *See also* death ray
development, 9, 21
Hoover on, 193
inspiring, 205
systems, 170
wireless power, 133
World System and, 182
wellhead, 129
Wells, H. G., 107, 159, 186, 195–97
Wells, Ida B., 102
Welles, Orson, 195
Westinghouse, George
Battle of Currents, 86–89, 104, 198
death of, 149
interest in designs, 84–86
letters to, 115
at Niagara Falls, 106
owning power, 122
partnership with, 90
proposals to, 132
shrinking capital, 98, 103
Westinghouse Electric Company
at Chicago World's Fair, 98
Niagara Falls contract, 105–7
relationship attempts, 173
royalty rights, 146
sale to, 115
working for, 90
World System letter to, 167–68
westinghoused, 87
White, Stanford, 110, 129–30, 141, 145–46
wireless power. *See also* Wardenclyffe Tower lab
with balloons, 147
electricity, 231
exploring, 3
mastering, 113
for national defense, 171
patents, 124
Poldhu Wireless Station, England, 143
protection, 4
system theory, 124
technology, 10
telegraphy, 142
transmitter, 87, 128–29
transportation, 132
universal, 242
weapons, 133
Wizard: The Life and Times of Nikola Tesla (Seifer), 207
wizard of electricity
balls of flame illustration, 94
at Chicago World's Fair, 95–96
creative breakthroughs, 112–15
Eggs of Columbus, 95–96, 117, 201
Madison Square Garden demonstrations, 113
polyphase project, 97–98
radiating electricity illustration, 97
social circle, 109–12
towers of power, 110
work ethic, 92–93
World Economic Forum (WEF), 241
World Magazine, 171
world peace, 5
World System
adopting, 190
funding, 10
letter, 167–68
promoting, 180
to protect democracy, 173
reclaiming, 243
vision, 185
weaponized, 182
world telegraphy system, 124, 129
World War I, 133
World War II, 193
worldview, 139, 230
Wotan, 121
Wright, Frank Lloyd, 102

X-rays, 10, 87–88

Yeats, William Butler, 184
Yetzer, Brian, 214

zero energy, 215, 242

ABOUT THE AUTHOR

John F. Wasik is a thinker, writer, poet, journalist, speaker, community activist, musician, and the author of sixteen books. He speaks around the world and contributes to the *New York Times, Forbes, CBS Moneywatch,* and newspapers on five continents. He lives in Grayslake, Illinois, with his wife and two daughters. www.johnwasik.net.

(No Model.)

N. TESLA.

ALTERNATING MOTOR.

No. 555,190. Patented Feb. 25, 1896.

Fig. 1

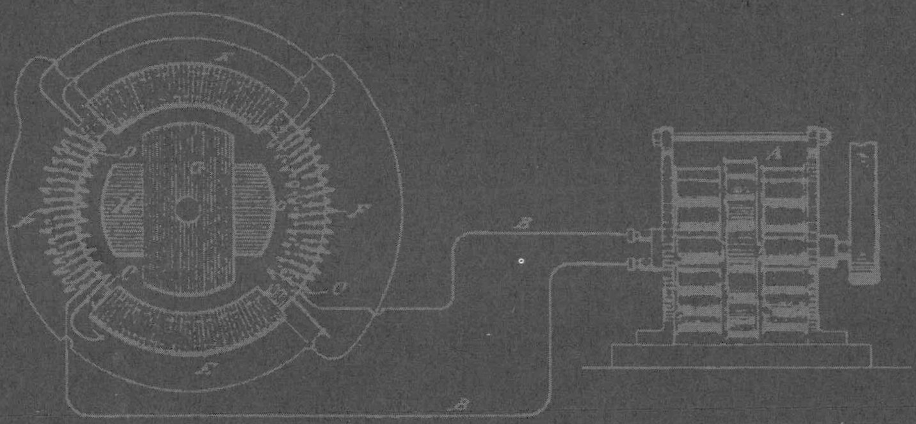

Fig. 2

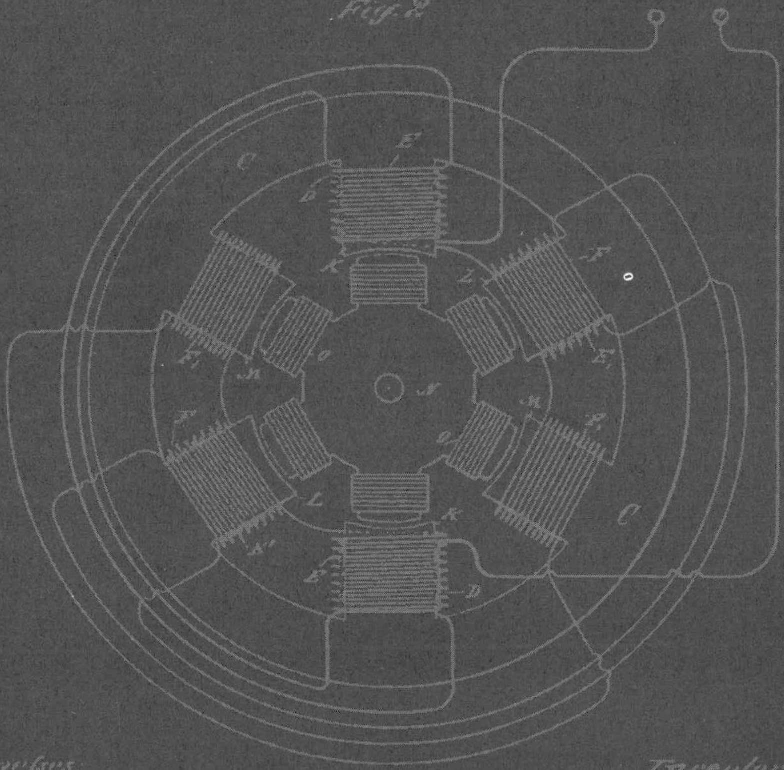

Witnesses:

Raphael Netter

Robt. F. Gaylord

Inventor

Nikola Tesla

by

Duncan, Curtis & Page

Attorneys.